Adriana G. Lopes, Andrew Brown
Single-Use Technology

Also of interest

Biodegradable Composites.
Materials, Manufacturing and Engineering
Kumar, Davim, 2019
ISBN 978-3-11-060203-6, e-ISBN 978-3-11-060369-9

Biosensors.
Fundamentals and Applications
Pandey, Malhotra, 2019
ISBN 978-3-11-063780-9, e-ISBN 978-3-11-064108-0

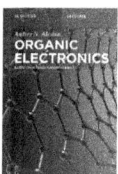

Organic Electronics.
Based on Hybrid Nanomaterials
Aleshin, 2020
ISBN 978-3-11-051846-7, e-ISBN 978-3-11-051850-4

Nanomaterials Safety.
Toxicity And Health Hazards
Ghosh, 2018
ISBN 978-3-11-057808-9, e-ISBN 978-3-11-057909-3

Bioelectrochemistry.
Design and Applications of
Biomaterials
Cosnier, 2019
ISBN 978-3-11-056898-1, e-ISBN 978-3-11-057052-6

Adriana G. Lopes,
Andrew Brown

Single-Use Technology

——

A Practical Guide to Design and Implementation

2nd Edition

DE GRUYTER

Author
Dr. Adriana G. Lopes
Biopharm Services
Unit 1 (1st floor)
Chess Business Park
Moor Road
Chesham HP5 1SD
United Kingdom
Adriana.g.lopes@gmail.com

Dr. Andrew Brown
Allergan Biologics Ltd
Estuary Commerce Park
Estuary Banks
Speke
Liverpool L24 8RB
United Kingdom
ucbeaib@gmail.com

ISBN 978-3-11-064055-7
e-ISBN (PDF) 978-3-11-064058-8
e-ISBN (EPUB) 978-3-11-064067-0

Library of Congress Control Number: 2019934969

Bibliographic information published by the Deutsche Nationalbibliothek
The Deutsche Nationalbibliothek lists this publication in the Deutsche Nationalbibliografie;
detailed bibliographic data are available on the Internet at http://dnb.dnb.de.

© 2019 Walter de Gruyter GmbH, Berlin/Boston
Typesetting: Integra Software Services Pvt. Ltd.
Printing and binding: CPI books GmbH, Leck
Cover image: Science Photo Library / LOOK AT SCIENCES / EURELIOS / PATRICE LATRON

www.degruyter.com

Preface

As the biopharmaceutical industry matures, trends towards increased flexibility and productivity, faster time to market and greater profitability are driving the replacement of traditional stainless-steel equipment by single-use technology (SUT). SUT use in the biopharmaceutical industry can impact the efficiency of manufacturing processes by reducing capital costs, improving plant turn-around-times, reducing start-up times and costs, eliminating non-value added process steps, and reducing the risk of cross-contamination. SUT has the potential to significantly reduce liquid waste, labour costs and on-site quality and validation requirements. SUT such as bags for preparation and storage of buffers have been used for many years to support bioprocessing activities. However, in recent years, SUT have been developed that can be applied to the majority of processing operations that can be found within the biopharmaceutical industry. For some of these operations, application of SUT is new and immature, which can present new challenges and risks to the end-user. In addition, there remain many unknowns regarding the way these technologies should be implemented into a validated, commercial GMP environment. This handbook aims to describe the activities that must be undertaken by the end-user to select and design technologies to meet requirements and implement SUT.

Chapter 1 starts by providing an introduction to SUT advantages, risks and the overall implementation process. Chapter 2 outlines recommended guidelines from regulatory agencies for adoption of a systematic science- and risk-based approach throughout the lifecycle of the development of medicinal drugs. Application of these approaches in the context of SUT implementation is discussed. This chapter also provides an overview of implementation plans, with emphasis on team structure and the use of risk-mitigation approaches. Chapter 3 describes considerations during SUT selection based upon a technical feasibility study and the business case of the system, as well as selection of a SUT supplier. Chapter 4 considers the specification and design of SUT systems. Chapter 5 underlines steps to validate a process using SUT. Chapter 6 presents case studies of key concepts applied to SUT technologies, such as: bag systems; bioreactors; tangential-flow filtration; formulation; and fill-finish. Here, considerations during the selection of off-the-shelf systems and custom-designed SUT assemblies are presented, whereby technical and business comparisons are made. Selection of appropriate material for a certain application and preliminary tests that must be undertaken are outlined. Examples of specification and design for each SUT system used for each application are presented, alongside process descriptions and flow diagrams. Risk assessments applicable to the design, processing and quality of the active pharmaceutical ingredient (API) are shown. Based on these assessments, qualification of the final SUT assemblies is presented. This handbook is the first comprehensive publication that describes the practical considerations that should be adopted at each stage of SUT implementation within biopharmaceutical facilities.

https://doi.org/10.1515/9783110640588-201

Contents

1 Introduction

Recent trends in the biopharmaceutical industry derived from technological advances (e.g., increased drug potency and smaller *niche* markets targeting patient-specific drugs) have resulted in the need for flexible manufacturing facilities. Further achievements in engineering cell lines capable of high production titres has led to a decrease in the volumetric manufacturing capacity needed to align with market requirements. Concurrently, the previous decade has seen the development of single-use technologies (SUT) applicable to biopharmaceutical manufacturing from the simplest and widely used bag systems and filters to more complex systems such as bioreactors, chromatography and fill-finish operations. As a result, industrial adoption of SUT has increased gradually, and end-users are considering application of the technology to operations discounted previously due to technical or scale limitations. Increased adoption of SUT has also brought about the realisation that new challenges are encountered to select, specify, implement and maintain the technology throughout the lifecycle of the active pharmaceutical ingredient (API). This handbook has been written to provide practical guidance on: (i) considerations for the end-user to review while choosing technologies to apply to processes; and (ii) implementation of SUT.

The route by which a process for the manufacture of biological products is designed, implemented and qualified can be long and complex. It requires the input of multi-disciplinary teams and there are many risks of failure. For example, the process can fail if it cannot be controlled to provide reliable batch-to-batch consistency of the product with sufficient quality. Exposure to risks by a particular organisations is dependent upon its experience with the product, process, manufacturing technologies and scale of operation. Many single-use systems are considered to be 'mature' because they have been present on the market for >10 years, been through design changes to improve performance, and have been utilised across a wide range of scales and manufacturing scenarios, from clinical through to commercial. However, other SUT are 'immature' and require more time to implement due to limited knowledge, availability and adaptability of the technology. There are no standardised approaches for SUT implementation. Instead, the implementation strategy should be 'tailored' based upon the type of technology and level of expertise of the end-user.

Compared with traditional stainless-steel systems, additional risks must be considered when using SUT. A comprehensive list of these risks is shown in Table 1.1. They have been grouped based upon impact to the end-user, supply chain, material and process.

These risks illustrate the range of capabilities that end-users must possess within their organisation, or that they will need to leverage from the supplier or third-part service providers to implement SUT. Hence, some end-users continue to employ traditional stainless-steel systems that they have expertise with, or adopt a

https://doi.org/10.1515/9783110640588-001

Table 1.1: Risks involved in adoption and use of SUT.

End-user risks	Supplier risks	Material risks	Process impact
Capabilities	**Business continuity**	**Material qualification**	**Quality**
– Materials and logistics	– Capacity	– Product compatibility and extractables	– Purity
– Stock management	– Single sourcing	– Migrants and leachables	– Contaminant profile
– Dual sourcing	– Disaster recovery	– Bioburden/sterility	– Product variants
– Bill of materials complexity	– Business continuity	– Particulates	– Location within process
	plans	– Animal-free components	
Knowledge of SUT		– Sterilisation	**Process performance**
– Experience and understanding of SUT	**Supplier quality**		– Titre
– Operator training	– Audit	**Material complexity**	– Yield
– Standardisation	– Change control and	– Compendial chemicals	– Throughput
– Testing new systems	end-user notification	– Integrated systems (standardisation,	– Control requirements
– Selection methodology	– Transparency and	dimensions, volumes)	
– Relationship with SUT supplier	compl- exity of the	– Integration with equipment hardware	**Facility fit**
	supply chain		– Available equipment
Knowledge of process		**Material integrity**	– Scale
– Understanding of the process/product	**Technical capability**	– Lot-to-lot consistency	– Local regulations
– Scale of operation	– Understanding of the	– Containment and integrity against leaks	– Support utilities required
– Equipment experience	process/product		– Warehouse requirements
	– Applications	**Handling**	– Flow of materials into
Quality	development	– Accumulation of particulates, endotoxin	and within the facility
– Supplier qualification	– Service and support	and bioburden contaminants	– Batch packing of
– Testing materials and release process		– Cleaning, disposal	consumables and
– Change control management		– Transport and storage	materials

'hybrid' approach whereby implementation of SUT is used to support process operations that use stainless-steel tanks (e.g., media/buffer preparation, hold, addition or waste collection). However, irrespective of whether the SUT is adopted fully or partially, the end-user should develop a robust implementation strategy so that risks are detected and mitigated in a timely manner.

There are many similarities between a project to implement a SUT and the traditional design approach for a stainless-steel system. However, there are differences, particularly in relation to the timing of when decisions are made and the criteria that are assessed. An overview of the key phases for implementation of SUT is laid out in Figure 1.1. An implementation plan should be developed at the start of the project and updated as progress is made, and a better understanding of the technology is developed. The end-user should start with assessment of the feasibility to use the SUT system for a given application, which should include technical and business assessments of the technology. Selection of an available supplier of SUT should be evaluated concurrently. Depending upon the complexity of the SUT for a given application, or maturity of the system, trial of a given system may be necessary before the technical feasibility can be completed. This strategy requires the co-operation of the SUT supplier, but should start with the end-user specifying the requirements of the system, including how it integrates with the wider process and facilities. At the end of the feasibility assessment, a decision to proceed with a preferred supplier is made. Hence, investigation of the quality and robustness of the supply chain of the supplier should be considered as part of this selection process. Once selected, the implementation plan should be updated when better understanding of regulatory acceptance of the SUT, system reliability and, above all, the resulting impact upon the quality of the product is known. A process-control strategy is required to ensure measurement of product quality, process interaction and validation. This strategy should underline the level of acceptable risk to the API in terms of cross-contamination, adsorption,

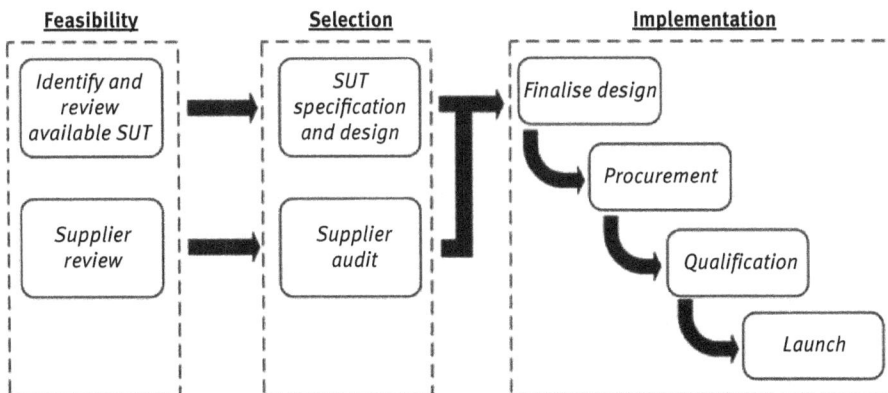

Figure 1.1: Key focus areas during SUT implementation.

and extractables/leachables from the SUT material that is product contacting, as well as process risks in terms of system integrity, process adjustments and operator safety. Specification, design and validation should ensure that the SUT system is fit for purpose so that, once implemented, it continues to support continued manufacture of the API to the required quality level. Once validated and in use, performance of the SUT should be monitored with metrics fed back to the supplier to ensure that issues are identified and dealt with in a timely manner.

1.1 Benefits and limitations of single-use technology

As the biopharmaceutical industry matures, the trends are towards the higher flexibility and responsiveness of production facilities as well as reduction of manufacturing costs and timelines in a background of increasingly strict regulatory and capacity demands. SUT can support an end-user to benefit from these trends but limitations do exist with the technology.

1.1.1 Improved process flexibility

By decoupling the process train from the facility infrastructure and transforming the facility into separate individual workstations it becomes easier to reconfigure the facility to meet changes in product scale or the type and number of products to be manufactured. The end result is greater flexibility with regard to the process and product. The portability of the equipment means that manufacturing spaces can be re-purposed as required. In addition, capacity can be increased through scale-up or scale-out, with minimal or zero impact to support systems such as water-for-injection (WFI) or generation of clean steam. As a result, SUT enables the drug manufacturer to increase manufacturing capacity and/or respond rapidly to market demands. If product demand increases, rapid expansion of capacity can be achieved by adding together similar SUT units with no need for implementation of process changes or improvements [1].

Single-use systems provide easier handling and quick turnaround times between batches and manufacturing campaigns due to the removal of clean-in-place (CIP), sterilisation and re-qualification activities [2]. This strategy improves process flexibility, and is particularly useful for multi-product facilities where assurance is required that the equipment is cleaned appropriately between batches of different products.

1.1.2 Increased speed of implementation

Faster construction, commissioning and launch of facilities can be achieved by using SUT. This is driven by the reduction in complexity of secondary support systems that would otherwise lead to longer design, fabrication and qualification activities. Single-use systems save time and money due to rapid product change-over and associated validation studies with minimal risk to product integrity, and results in accelerated time to market [3]. It also means that capital decisions can be delayed without impacting timelines for drug development. This approach reduces the risk that a decision to build a facility is taken when the capacity required is unclear or likely to change. If a manufacturer of a drug for clinical trials requires to build a clinical facility, SUT is much faster to implement than a traditional stainless-steel facility. Also, the overall costs of implementation are lower so, if the drug fails clinical trials, it carries a reduced risk to the business due to the flexibility of re-configuration to a new product and reduced capital costs.

1.1.3 Cost savings

sually, single-use systems are supplied pre-sterilised (by gamma radiation), thereby eliminating the need for CIP or steam-in-place (SIP) support systems, areas and procedures, as well as the equipment maintenance associated with these practices [4]. Reduction of capital investment costs for process equipment is achieved by elimination of utility requirements for CIP and SIP capabilities, and reduction of the number and size of CIP skids [2, 5]. Due to elimination or reduction of CIP and SIP requirements, generation of purified water (PW) can also be reduced in scale and cost for new-build facilities.

SUT also results in better utilisation of facility assets. The reduced scale of SUT equipment (smaller facility footprint) results in reduced fixed costs (e.g. investment, operation, maintenance) and a 'better utilised facility' that can respond to higher demands in production by process intensification.

1.1.4 Increased product safety

Single-use operations result in a reduced risk of cross-contamination and increased assurance of sterility [6] due to elimination of cleaning between batches and the associated validation. The low detection limit assays used to measure contaminants after cleaning, combined with the lack of acceptable cross-contamination levels, increase the risk associated with cleaning procedures. SUT systems are used only once for a specific process and operate in a closed-system environment, which prevents cross- contamination of product and protects operators. A closed system also allows different operations to be undertaken concurrently in the same room while minimising the impact on heating ventilation and air conditioning (HVAC) airflows and pressure differentials.

1.1.5 Technical limitations

The main limitations of SUT are based on the scale of operation as well as the ease of scalability and operability. Available bioreactors using disposable technology may reach only ≤4,300 l (working volume, 3,500 l), and disposable chromatography columns have diameters of ≤60 cm. Scale limitations are typically due to the strength and durability of the plastic material. In general, SUT are not recommended if they are likely to come into contact with organic solutions, or in operations requiring high heat removal transfer or high mixing rates.

Some SUT provide scalable options but the end-user would have to use the same system and supplier. Scaling up or down between different technologies is more difficult due to the lack of inter-changeability between them as well as different system designs and configurations. Sometimes, unconventional and unproven scale-up/down methodologies must be considered [7]. Finally, the process performance of a SUT system may not have been proven completely compared with the traditional stainless-steel equipment it is intended to replace. Main concerns involve the ability to deliver similar mixing, pressure and flow rate, as well as control capability to deliver a process and product reproducibly and consistently.

1.1.6 Cost increases

Use of SUT leads to increased operational costs resulting from repeated use of consumables or items that would otherwise be manufactured from stainless steel. If items are used once per batch then there are also increased costs derived from waste disposal, which need to be managed internally. Depending upon the number required, cost and availability of single-use items, these may become the drivers of cost of goods. Facilities with high use of SUT have an added emphasis on logistics and workflows resulting from the changed requirements of storage and manual transport of process liquids, equipment, consumables and waste, as well as redesigned personnel flows [8, 9].

1.1.7 Increased complexity

Lack of maturity of some SUT systems (and associated operational experience) poses new challenges and risks. Ease of use may not be proven fully within a manufacturing environment, or the robustness of the system may not be known from clinical to commercial scales. The number of SUT systems available for a particular application may be limited. Lack of standardisation across suppliers in the utilisation of materials and connections as well as integration with hardware also increases the amount of design and review work required, and reduces the ability of the end-user to identify secondary source suppliers. As a result, there are limited options for inter-changeability and connectivity between similar technologies.

There is a lack of guidelines and standard procedures for the use and validation of SUT. Use of SUT introduces new requirements for validation of plastic product contact materials, such as integrity, sterility, and compatibility with the product. An example of this absence is the test conditions used for assessment of extractables, where standard methodology is lacking [10].

Complexity can also arise from the tubing arrangements, assembly and disassembly of single-use components, operation of sterile connections between equipment and components, and steps required to achieve a leak-free environment. Design approaches can be taken to reduce this complexity, particularly if a SUT is implemented across an entire process. However, a new layout of the facility, work flows and training approaches are required so that operational handling errors are minimised.

1.1.8 Dependence on suppliers

A major concern for SUT use is the dependence on suppliers to provide a consistent and cost-effective supply of systems that meet the required quality specifications [11]. SUT require repetitive purchases, and suppliers must be certain that they have sufficient capacity to ensure a consistent supply of single-use components. In turn, the end-user must have detailed understanding of the supply chain of the SUT and must consider inventory management and storage capability.

As mentioned above, the limited availability of components/parts and restricted interconnectivity between different technology/suppliers places considerable emphasis on selection of the appropriate supplier and materials provided. Selection of suppliers and qualification of the vendor's quality systems becomes very important to ensure robustness of the supply chain.

The potential advantages of SUT presented above can make a compelling case for adoption. To minimise risk, the disadvantages should form the basis of the considerations that need to be evaluated during the selection and implementation of this type of technology.

References

[1]   C. Valle, *Filtration Separations*, 2009, **46**, 18.
[2]   A. Sinclair and M. Monge, *Pharmaceutical Engineering*, 2002, **22**, 1.
[3]   T. Kapp, *BioProcess International*, 2010, **8**, S10.
[4]   Pall Corporation, GDS Publishing Ltd., Bristol, UK, 2011. [Private Communication].
[5]   A. Sinclair and M. Monge, *BioProcess International*, 2011, **9**, 12.

[6] J. Robinson and B. Bader in *Proceedings of the Interphex Conference & Exhibition 2008*, Pennsylvania Convention Center, PA, USA, 2008, p.1.

[7] R. Eibl, S. Werner and D. Eibl, *Advances in Biochemical Engineering/ Biotechnology*, 2009, **115**, 55.

[8] N. Guldager, *Pharmaceutical Technology*, 2009, **33**, 68.

[9] M. Monge, *BioPharm International*, 2006, S43–S51.

[10] A.G. Lopes, *Food and Bioproducts Processing*, 2015, **93**, 98.

[11] A. Ravise, E. Cameau, G. De Abreu and A. Pralong, *Advances in Biochemical Engineering/ Biotechnology*, 2009, **115**, 185.

2 Strategies for implementation of single-use technology: A risk- and science-based approach

The amount of resources and effort required when implementing single-use technologies (SUT) are dependent upon the scope of application and risk factors identified in Table 1.1. If application is in a manufacturing environment, then understanding the impact upon product quality is important. Also, the ease of use and quality of single-use components must be acceptable. If the application is late-stage clinical or commercial manufacturing, then the robustness and reliability of the supply chain will be important factors. If the end-user has little experience in working with SUT, or is investigating use of SUT for a new application, then effort will be required to assess and test the SUT to ensure that it is suitable. If the SUT is being considered for a single-unit operation, then implementation will require less effort than if the SUT is being applied throughout an entire process or facility.

An approach for SUT implementation is presented in Figure 2.1. The degree to which the approach is applied varies depending upon the scope of application. There are three key phases to implementation. First is the initial assessment of the feasibility of the SUT to an application. The key milestone reached at the end of this phase is identification of feasible options to proceed and investigate further. The second phase is the design and selection of the SUT from a supplier. The milestone at the end of this stage is sanction of capital to implement the SUT selected. During this phase the design is finalised, and systems are procured, qualified and launched. The goal of this phase is the release of a single-use system for use within a manufacturing process.

Various aspects of the implementation approach will be explored throughout this handbook and the relevant sections are indicated within Figure 2.1. Key concepts of the approach are:

1. An implementation plan is developed that addresses the requirements of all stakeholders and considers (from the beginning) the criteria required for successful implementation;
2. A multidisciplinary team should be assembled with knowledge of the process in addition to operations, quality, engineering and logistics;
3. Design considerations should be considered early in the implementation approach, and the preferred technology should be tested at scale by operational personnel before making purchasing decisions; and
4. A systematic science- and risk-based approach should be adopted to identify and mitigate risks reviewed throughout the implementation approach. Risk areas considered should include technical, business, quality and supply chain.

https://doi.org/10.1515/9783110640588-002

Feasibility → Feasibile options exits to proceed → **Selection** → Capital Sanction → **Implementation** → Ready for use

Implementation Plan (Section 2.2)
- Scope of implementation
- Application (R&D, GMP, lab, pilot, etc.)
- Project plan broken down by phases
- Team & responsibilities
- Quality activities

Technical (Section 3.1)
- Manufacturing strategy
- Determine key process or orperational requirements
- Identify specific SUT that will respond to requirements
- Identify Key staff that will be using the technology and who will be required to implement it successfully

Business case (Section 3.2)
- Business drivers
- SWOT assessment
- Capital and operational cost implication vs. alternative options such as stainless steel or existing facility capabilities
- Impact upon facility and existing operations
- Quality and logistical implications

Supplier assessment (Section 3.3)
- Review available suppliers
- Supplier questionnaires
- *Case studies presented in Chapter 6*

Trial (Section 3.1)*
- Hands-on trial with technology at required scale, either on-site or at suppliers facility
- Test under process conditions /model system

Concept Design (Chapter 4)
- URS, process and SUT component drawings
- Preliminary equipment list
- Selection of preferred technology
- Materials (BOM) and equipmnet list
- Warehouse and logistical controls
- Flow of materials and waste within facility
- Capital cost
- Operational cost forecast
- Technical and quality risk assessment

Supplier review (Section 3.3)
- Supplier risk assessment
- Supplier audit
- Supplier management strategy, draft supply chain & quality agreements
- Identify possible dual source supplier/technology

Single-use Extractable Review (Section 5.2)*
- Extractable risk assessment
- Review extractable data package where available

Update Implementation plan

Finalise Design (Chapter 4)
- URS, process and SUT component drawings
- BOM, Equipment list
- Testing plan and validation protocols

Material Qualification (Section 5.1)
- Leachable assessments where applicable
- Integrity
- Compatibility
- Sterility and cleanliness

Qualification and validation (Chapter 5)*
- Equipment FAT, SAT and QA approval
- Installation & operational qualification
- Maintenance and calibration documentation and frequency
- Staff training
- Generation of manufacturing documentation
- Operator and QC training
- Performance qualification

Launch (Section 5.3)
- Complete qualification of suppliers
- Finalise & execute supply chain & quality agreements
- Approval of all qualification documentation
- Write material specifications
- Build consumable stock for launch

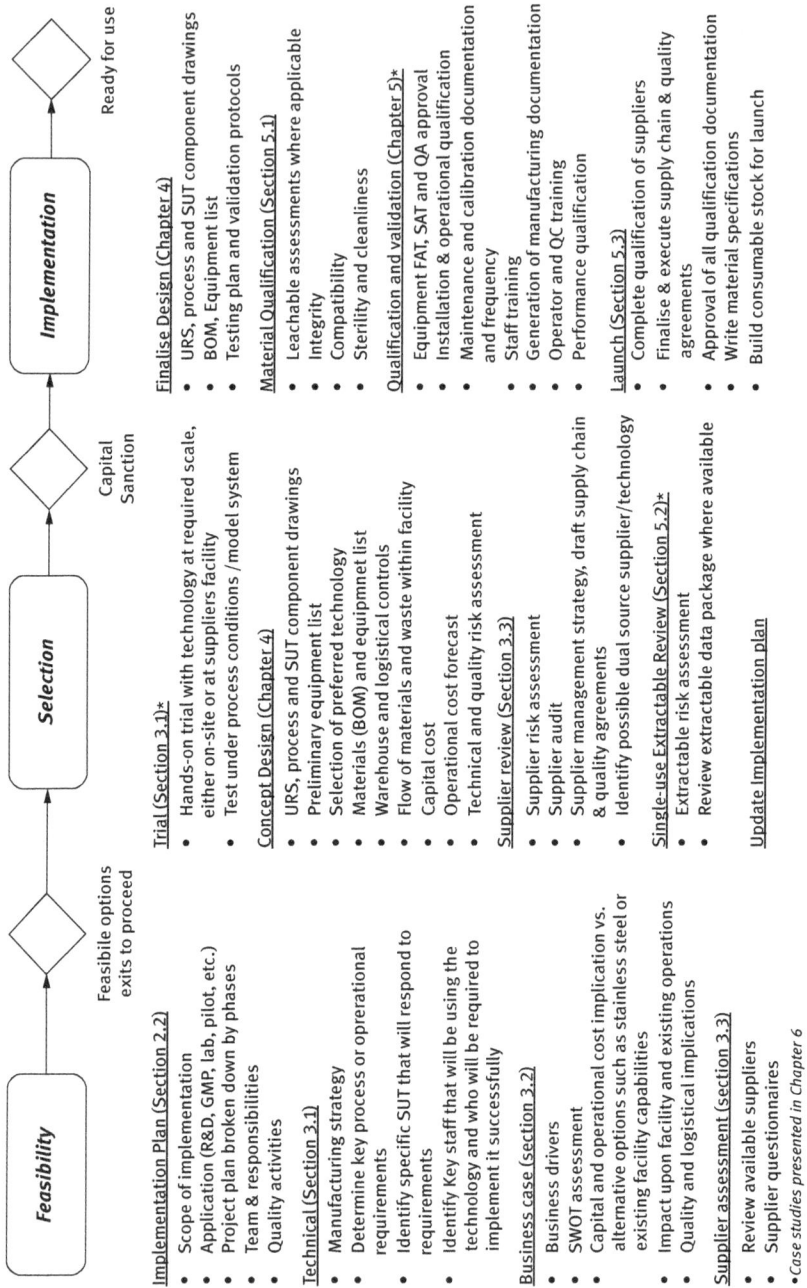

Figure 2.1: Recommended approach for SUT implementation. BOM: Bill of materials. QA: quality assurance; QC: quality control; R&D: research and development; SAT: site acceptance testing; SWOT: strength, weakness, opportunities and threats analysis; and URS: user requirements specification.

2.1 A risk- and science-based approach

The regulatory bodies that oversee development and manufacture of therapeutic drugs have numerous published guidelines that guide companies in their activities. Many of the principles covered in these guidelines are applicable to the selection and implementation of new technologies. Key aspects are to build quality and undertake risk assessments throughout the implementation process to ensure that the technology works as intended and does not impact upon the active pharmaceutical ingredient (API) quality. Current good manufacturing practices (cGMP) guidelines for pharmaceuticals set by the Food and Drug Administration (FDA) require the industry *'to integrate quality systems and risk management approaches into its existing programs...'* and state that *'quality should be built into the product'* from the development phase and throughout the lifecycle of a product [1]. *'The FDA has identified a risk-based orientation as one of the driving principles of cGMP initiative ... The goal is to use a scientific framework to find ways to mitigate risk while facilitating continuous improvement and innovation in pharmaceutical manufacturing in the context of risk- and science-based approach'* [2].

GMP guidance from the FDA states that industry should use technologies that facilitate conformance with cGMP and streamline product development [2]. In the case of SUT, they aid conformance with GMP and can streamline operations through:

1. Reduced cleaning and potential for contamination;
2. Dedicated equipment and/or disposable parts;
3. Simpler change-over between products in multi-product facilities; and
4. Use of closed process equipment to alleviate the need for stricter classification of rooms.

To integrate quality systems and risk-management approaches, the International Conference on Harmonisation (ICH) has established quality standards and requirements. The relevant guidelines are:

- ICH Q8 *'Pharmaceutical Development'* [3] incorporates elements of risk and quality by design;
- ICH Q9 *'Quality Risk Management'* (QRM) [4] relates to quality and GMP compliance; and
- ICH Q10 *'Pharmaceutical Quality System'* [5] covers the lifecycle management of process and system control.

The outcome of the risk-management framework is intended to lead to a science-based decision undertaken across the lifecycle of a product. In the case of a SUT, because it is not fully mature, several sources of risk arise, particularly those derived from material/human failure or a lack of knowledge of working with such technology. These risks are presented in Table 1.1 and should be addressed by an end-user as they seek to adopt SUT.

ICH Q8 presents the concept that quality should be built into the drug product from the beginning (i.e., starting with the design process) and is, therefore, not solely reliant upon retrospective testing of product or intermediates. This strategy relates to manufacturing systems in that a 'quality by design' approach is used to ensure that critical aspects are designed into systems during the specification and design process, and are documented alongside their acceptance criteria in a suitable manner. Assurance that manufacturing systems are fit for intended use [e.g., material attributes or control of critical process parameters (CPP)] should be monitored and evaluated continuously throughout the lifecycle of the system. API and process information as it relates to drug product quality and patient safety, should be used as the basis for making science- and risk-based decisions. This approach is seen as a means to ensure that manufacturing systems are designed and verified to be fit for their intended use.

API and process information can come from many routes: scientific investigation; previous manufacturing experience; understanding of regulatory frameworks and their applications. A company may already have considerable API and process understanding to leverage depending upon the maturity of the drug product. Regardless of the initial level of experience, knowledge will increase as the end-user progresses through the design, verification and implementation process. This knowledge should be captured and communicated so that it can be used effectively and decisions re-assessed.

API and process information that should be considered or analysed further includes critical quality attributes (CQA) of the API, CPP, and information on the process control strategy. The definition of CQA and CPP as stated by ICH Q8 [3] are:
- CQA is a physical, chemical, biological or microbiological property or characteristic that should be within an appropriate limit, range, or distribution to ensure the desired product quality; and
- CPP is a process parameter whose variability has an impact on a CQA and, therefore, should be monitored or controlled to ensure the process produces the desired quality.

The SUT may in its operations have a direct or indirect impact upon CQA. Risk- assessment tools should be used to identify the level of impact, and ensure that suitable steps are identified and taken to mitigate the risk during design and implementation phases. The design and implementation process can be considered to be successful if the equipment and facilities with the corresponding control systems achieve the requirements for CPP or CQA and eliminate (or control appropriately) risk to the patients. Verification activities after implementation should define acceptance criteria based on these critical aspects and should be documented. If manufacturing systems meet the required criteria, then they show evidence that they are fit for the intended use. A focus upon CQA of manufacturing systems should lead to efficient use of resources, but verification inspection and tests should not be limited to only critical aspects.

In accordance with ICH Q9, risk management should underpin the specification design and verification process with the focus on risk to product quality and patient safety. '*The evaluation of the risk to quality should be based on scientific knowledge and link to the protection of the patient. The level of effort, formality and documentation of the quality risk management process will depend upon the level of risk to product quality and patient safety*' [4]. Quality risk management (QRM) is a systematic process for the assessment, control, communication and review of risks to the quality of the (medicinal) drug product across the product lifecycle. It supports a scientific and practical approach to decision-making. QRM provides documented, transparent and reproducible methods to accomplish steps of the QRM process based on current knowledge about assessing the probability, severity and (sometimes) delectability of the risk. Organising data and facilitating decision-making through cause-and-effect diagrams (also called Ishikawa or Fishbone diagrams) and failure mode effects analysis (FMEA) provides an evaluation of potential failure modes for processes and their likely effect on outcomes and/or CPP and CQA. Higher risks should have a higher level of control and documentation. Risk-management tools can also be utilised to assess and control the robustness and performance of the SUT and the quality of the supply chain.

The QRM process consists of a series of steps that begin with initiation of the process, in which the scope is defined. A risk assessment is conducted to identify hazards, and to analyse and evaluate the risks associated with these hazards. Risk related to the design and implementation of manufacturing systems includes the impact of technological novelty or complexity in addition to vendor/material risks upon product quality and patient safety. If risks cannot be eliminated, then controls are considered to reduce the risk to a level whereby they are acceptable. The result of risk assessment is communicated out to the wider organisation, and is reviewed on a periodic basis to ascertain if events have impacted upon the evaluation and acceptability of the risk.

Use of API and process information as well as QRM enables implementation of ICH Q10 effectively and successfully across all stages of the drug product lifecycle [5].

ICH Q10 facilitates innovation and continuous improvement of process performance, API quality, the QRM system, and strengthens the link between development and manufacturing activities. The objective of the quality system is to maintain a system that delivers product with appropriate quality attributes consistently. This objective is achieved by using QRM to monitor and control systems, process performance, and product quality. In the case of SUT systems, the quality system extends to the control and review of outsourced activities and quality of purchased materials (including management of responsibilities). Ultimately, the end-user company is responsible for ensuring that processes are in place to assure control of outsourced activities and the quality of purchased materials.

Using a science- and risk-based approach for SUT implementation is paramount for identification, quantification and management of the critical sources of variability that may affect API quality. Control strategies can be used to maintain a state of control and facilitate continual improvement applied throughout the product lifecycle.

2.2 Implementation plan

Outlining a project plan with detailed timescales, responsibilities of individual stakeholders, and deliverables is paramount to the success of the SUT implementation strategy. The project manager (end-user) should hold implementation meetings to determine levels of support available from each stakeholder and to identify the critical activities to be undertaken during the course of the project. Goals and milestones should be defined, and teams and responsibilities should be delegated as required to ensure timely delivery of tasks. An implementation checklist is shown below:

- Describe goals and objectives.
- Identify the roles and responsibilities of stakeholders (Figure 2.2).
- Identify impacts/bottlenecks and methods to overcome or mitigate.
- Identify resources that are needed and if they are available, the desired completion date, and constraints.
- Subdivide the implementation plan into steps (Figure 2.1).
- Identify key milestones/decision points to be tracked.
- Project paths and methods of progression tracking.
- Schedule team meetings.

The implementation process is a team effort involving stakeholders from procurement (supply chain), planning, operations or manufacturing science and technology (MSAT), process engineering, quality, and the SUT supplier. Subject matter experts (SME) should take the lead role in verification that, based upon their area of expertise and responsibility, the manufacturing systems are appropriate. Responsibilities include defining verification strategies, acceptance criteria, selection and execution of appropriate test methods, and review of results. A strong collaboration with the SUT supplier is advisable, but may not be required if the organisation has good experience with the SUT for the scale and purpose that it is to be applied. Otherwise, the SUT supplier should be part of the project team to provide expertise in SUT technology, along with help with the integration into existing technologies, operations and systems. Figure 2.2 lists the project team and a summary of its tasks and responsibilities during SUT implementation. The make-up of the team will

```
                                ┌─ Suppliers audit
                                │
                                ├─ Materials qualification
                 ┌─ Quality ────┤
                 │              ├─ Review of all associated manufacturing documentation
                 │              │
                 │              └─ Setting up and maintenance of quality procedures and supervision of
                 │                 validation activities
                 │
                 │              ┌─ Setting up overall procurement strategy
                 │              │
                 ├─ Procurement ┼─ Development of a secure supplier strategy with emphasis on a second
                 │              │  alternative supplier
                 │              │
                 │              └─ Providing analysis of capital and procurement costs
                 │
                 │              ┌─ Inspect or test equipment against engineering specifications, and
                 │              │  investigating departures from specifications
                 │              │
                 │              ├─ Identifying the best technology (together with operations/MSAT)
                 ├─ Engineering ┤
                 │              ├─ Undertake the detailed design to meet procurement cost and quality
                 │              │  requirements
                 │              │
                 │              └─ Support validation activities
Project Manager ─┤
                 │              ┌─ Support technology trial/evaluation of SUT
                 │              │
                 │              ├─ Definition of the URS
                 │              │
                 │              ├─ Identification of operational risks and considers operational compromises
                 │              │  associated with the use of a particular SUT
                 │              │
                 ├─ Operations  ┼─ Determine CQA and CPP
                 │  or MSAT     │
                 │              ├─ Process scientists to lead the resolution of process and product related issues
                 │              │  during scale-up and technical transfer
                 │              │
                 │              ├─ Manufacturing personnel to review operability of the design in addition to the
                 │              │  development of documentation such as standard operating procedures
                 │              │
                 │              ├─ Validation support
                 │              │
                 │              └─ Training staff
                 │
                 │              ┌─ Suppot all stakeholders with information required for each stage such as
                 │              │  certification of technology components, auditing
                 │              │
                 └─ SUT supplier ┼─ Generate protocols and support or conduct validation activities
                                │
                                └─ Support training in SUT
```

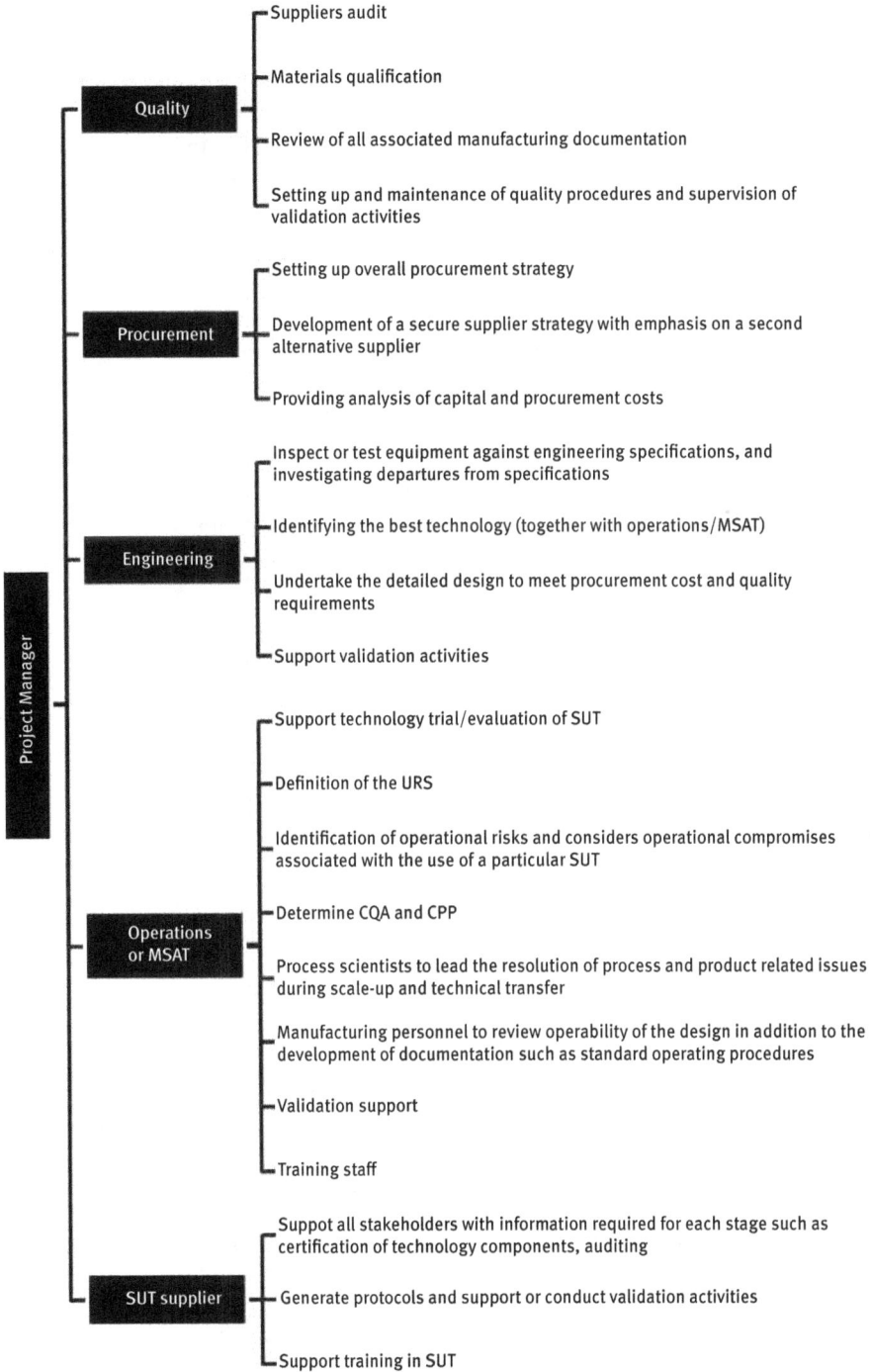

Figure 2.2: Members and responsibilities of the implementation project team.

probably change throughout the project because not all SME are required from the beginning of the project.

Implementation of SUT starts with factors that determine if application of the SUT system for a given application is feasible. Here, assessment of technology and the business case is drawn (more details are given in Sections 3.1 and 3.2). Criteria for selection and evaluation of a particular SUT are listed. If any of the high-level criteria are not met then single-use systems for the envisaged application may not be applicable. The risk-based strategy starts when choosing which SUT to implement. After initial definition of the process, a supplier is selected and SUT components or assemblies identified. The importance of ensuring compatibility of equipment and material is paramount for ensuring the quality of the final drug product. Equally important is to undertake initial assessment of the manufacturing process, quality systems and sourcing of materials when choosing the SUT supplier. A risk-based approach should be applied when evaluating the security of supply and qualification of SUT suppliers. Some of the risk-mitigation strategies concern identification of alternative vendors and alternative options for disposable parts. An alternative and preferred SUT options can be tested concurrently to reduce risk.

A risk-based approach should also be applied during the design, qualification and continued verification steps. During the design stage, the project team is responsible for undertaking a risk assessment to identify CPP for SUT operations and consider the impact on API characteristics before the validation process. A risk assessment of all individual single-use components, assemblies and support equipment is undertaken at this stage to demonstrate no risk or a risk-mitigation strategy to drug product quality and patient safety. Outcomes of the risk assessment form the basis of a process validation approach that ensures that manufacturing systems are fit for purpose when implemented, and that they will continue to support continued manufacture of the drug product.

Validation and qualification studies must demonstrate the suitability of the SUT system for the end application. Examples include:
- API compatibility
- Studies on microbial control and impurities
- Hold studies to establish acceptable product/intermediates hold durations
- Growth and yield of cells (in the case of culture bags)
- System integrity
- Testing of leachables

Validation requirements are justified based upon risk-based activities such as criticality and impact assessments. These identify critical and non-critical systems and their components; and then evaluate whether the systems have a direct impact on the product quality, purity, safety and effectiveness.

Once the process and system is qualified, then continued process verification is required. The aim of this activity is to monitor and control performance, including process changes, vendor change notifications and undertaking periodic re-qualification and maintenance. Risk analysis is also useful during this verification stage because it aids identification of failure in operating parameters that fall out of a proven acceptable range or, in a worst-case scenario, upon process/system robustness. Risk analysis ensures that controls are in place to alert users if process variables start to fall out of control.

2.3 Risk-assessment tools to support implementation

Risk assessment consists of identification of hazards followed by the analysis and evaluation of risks associated with exposure to those hazards. Risk is a function of two contributing factors: probability of occurrence and severity of harm. The higher the two factors, the higher is the overall risk.

Various risk-assessment tools are designed to support a scientific approach to decision-making. Common examples include checklists, flowcharts, process maps, cause-and-effect analysis (Fishbone) diagrams, preliminary risk assessments (PRA) and control charts. Other more formal tools referred to in ICH Q9 are: FMEA; hazard analysis and critical control point (HACCP); hazard and operability studies (HAZOP). These methodologies should be selected according to their suitability for a particular application. For example, the HACCP is more suitable when the drug product has been launched because it facilitates monitoring of critical points in the manufacturing process. The HAZOP is usually undertaken during the design stage to evaluate process safety hazards and deviations from normal use (original design intent).

The overall process of risk management is described in ICH Q9. It starts with identification of the risks and hazards associated with each step of process. A PRA can be used at the earliest stage in the design process to identify risks if uncertainty remains and few parameters have been defined [6]. In effect, it is a 'brainstorm' of 'what if' scenarios and alternative design options. The cause-and-effect analysis of each individual hazard identified can be visualised in a Fishbone diagram. The latter presents the product or process in the main 'spine', and the secondary 'spines' are different factors or causes of hazards. The hazard can be derived from failures in materials, controls, personnel, equipment and procedures. Once the risk and source have been identified, assessment of the impact of potential failures can be undertaken and quantified. The impact or criticality of the risk can be assessed by measuring its severity and probability of occurrence against the number of controls required to eliminate or reduce risk to an acceptable level. This analysis can be undertaken in a FMEA, which is a tool used to: identify potential failures; examine their impact upon product quality; propose adequate corrective and preventive actions. The FMEA includes quantification of the following aspects related to risk:

- Severity (S), or how significant the deviation is in terms of product quality and patient safety;
- Occurrence (O), probability and frequency of occurrence; and
- Detectability (D), which includes controls and methods to detect deviations (or their associated parameters).

The severity, occurrence and detectability of risk are multiplied to obtain a risk priority number (RPN) that is used to differentiate which areas carry more risk. The higher the RPN, the greater the need of additional controls and/or more frequent requalification processes than others.

Based upon FMEA results, a control strategy for the identified hazards should be implemented, monitored and reviewed continuously. The FMEA can be used to assess risk at different stages of SUT implementation. It can be undertaken during design of a single-use final assembly or equipment to assess the risk of failure of components or assembly method. The FMEA can also focus on process failures (CPP) to ensure reproducibility between batches or the product quality (whereby the impact upon CQA is the main focus). The FMEA can, therefore, aid determination of which critical processes require design verification and process validation. Examples of risk assessments applied to all stages of SUT implementation and specific to different bioprocess operations (e.g., upstream, downstream, fill-finish operations and process support systems) are presented in Chapter 6.

References

[1] *Quality Systems Approach to Pharmaceutical cGMP Regulations*, Department of Health and Human Services, Food and Drug Administration, Rockville, MD, USA, September 2006.
[2] *Pharmaceutical cGMPs for the 21st Century – A Risk-based Approach*, Department of Health and Human Services, Food and Drug Administration, Rockville, MD, USA, September 2004.
[3] *International Conference on Harmonisation of Technical Requirements for Registration of Pharmaceuticals for Human Use, Pharmaceutical Development Q8(R2)*, ICH Harmonised Tripartite Guideline, Step 4 Version, International Conference on Harmonisation, Geneva, Switzerland, August 2009.
[4] *International Conference on Harmonisation of Technical Requirements for Registration of Pharmaceuticals for Human Use, Quality Risk Management Q9*, ICH Harmonised Tripartite Guideline, Step 4 Version, International Conference on Harmonisation, Geneva, Switzerland, November 2005.
[5] *International Conference on Harmonisation of Technical Requirements for Registration of Pharmaceuticals for Human Use, Pharmaceutical Quality System Q10*, ICH Harmonised Tripartite Guideline, Step 4 Version, International Conference on Harmonisation, Geneva, Switzerland, June 2008.
[6] J. Vesper in *Risk Assessment and Risk Management in the Pharmaceutical Industry*, Parenteral Drug Association, Bethesda, MD, USA, 2006.

3 Feasibility assessment of single-use technology and suppliers

Selection of single-use technologies (SUT) should first focus on determination of whether application of a SUT for a given application is feasible. This determination should include a technical and business assessment of the technology in addition to review of the available suppliers and their capabilities. The first step of this evaluation is assessment of whether the technology will meet process requirements such as operation, product yields and quality. Limitations in terms of scale or technical performance that could impact upon the required outputs should be considered. The cost of investment and the cost derived from operation of the system should consider scale, infrastructure requirements (support systems and utilities), regulatory requirements [biosafety, good manufacturing practices (GMP), room grades] and the experience and training of personnel. The logistical control strategy impacts upon facility requirements in terms of storage and material release. In addition, a strategy to guarantee security of supply and qualified vendors is very important. When choosing a SUT, the supplier will become a business partner: this is a key decision and should be included as a part of the system selection from the early stages. Some of the considerations when choosing suppliers are: the service support provided to clients; quality of SUT supplied; and track record of continuous supply of SUT of appropriate quality.

3.1 Technical feasibility

Not all SUT are mature, and even those that have been utilised for many years and across a wide range of applications may not fit the specific application of end-use. In all instances where the application is new to the end-user, thorough technical evaluation should be carried out. The first step of this technical feasibility analysis involves preliminary assessment of the types of SUT available and the suitability of system to achieve the intended aim. Table 3.1 summarises the technical criteria related to process, operations, facility and the SUT that should be considered during this stage.

The technical criteria during processing should address the suitability of the plastic material under normal operating conditions (volume, pH, temperature, pressure, flow rate) and the resulting impact on process, product yield and quality. The sterility of single-use components is important because the method of sterilisation can affect material properties and require additional validation.

Scale of operation and scalability of the system should be assessed to ensure that the limits do not impact upon the end-use. For example, single-use bioreactors (SUB) are available in discrete sizes up to a scale of 4,300 l (working volume, 3,200 l).

https://doi.org/10.1515/9783110640588-003

Table 3.1: Technical criteria to be reviewed during assessment of the technical feasibility of SUT.

Process (Figure 3.1)	– Materials of construction and physical properties (biocompatibility and leachables) – Scale of operation/size – Physical attributes and process parameters (time, pressure, temperature, volume, pH, mixing, flow rates) – Sterility required? – Closed system required?
Operations	– Flexibility/complexity – Set-up requirements – Robustness – Connectivity with other operations or support equipment – Calibration – Integration with hardware and control system – Monitoring and sampling of the process – Personnel/resources requirements – Support systems (air, water, welder, fuser) – Safety
Facility	– Room classification – Storage of SUT (warehouse, within processing rooms, quality release) – Utility requirements (steam, high-purity water) – Floor space, ceiling height – Procurement (purchase orders, SUT type, quantities)
SUT	– Off-the-shelf* system – Standard supplier designed system with configurations – User-defined system – Maturity of the technology

* Pre-designed, pre-tested and completely integrated for an application

The connector size typically supports a maximum diameter of ¾ inch and is suitable for tube welding with a pump flow and pressure rating <3–5 bar gauge [1]. Gas flow and bubble size into the SUB is determined by the number and type of spargers available. The number and type of the internal mixer design is limited according to how much power can be inputted into the system. These technical characteristics must be cross-referenced against the known or estimated requirements of the process to ascertain if the system can support the application. For example, the ability to monitor critical process parameters (CPP) through robust systems of online sensors or offline *via* sampling is fundamental. Choice of sensors should be based on their robustness for use within GMP operations [1].

For bioreactor applications, the bag design defines whether conventional dissolved oxygen (DO) and pH sensors can be used instead of new single-use optical sensors. Conventional stainless-steel sensors require validation of the sterilisation method, but they may be of benefit to the end-user due to their greater accuracy over the course of the operation than available single-use sensors. In addition to being compatible with bag ports, sensors must also be compatible with the associated control systems.

More detail on the specification and selection of a system for cell culture can be found in Case Study 2 in Section 6.2.

The use of the SUT impacts upon the design of a facility. For an existing facility, considerations should include the impact on installed equipment and processing steps. Depending upon the existing design this may require varying degrees of modifications because the facility may require:

- Re-design of material, personnel and waste flows;
- Re-configuration of utility requirements;
- Additional space, such as the floor footprint or ceiling height;
- Extension of storage capability; and
- Room re-classification and changes in room configuration, which will depend upon the operations that are impacted by the SUT. This parameter could impact upon heating ventilation air conditioning (HVAC) systems.

For some single-use systems there may not yet be a body of reference data spanning many years of operation with different types of systems and scale of manufacture. In these cases the technology is considered to be immature. The end-user should identify whether the technology has been tested with a similar cell line, product type, material stream, or under equivalent process conditions [2]. Such identification can often present a challenge, with little valid data available from the literature or supplier sources. To combat this lack of data, and if the SUT is critical to operations, then a trial of the technology should be undertaken by the end-user so that a dataset can be obtained. If incorporated with risk assessment, this strategy can help the end-user to build confidence of the suitability of a particular technology to process needs. If alternative technologies are available then they should be compared. Compatibility with the other technologies that will interact with the SUT should be considered. Depending upon the process, these may be single-use or conventional stainless-steel systems.

Figure 3.1 presents an example of a typical biopharmaceutical process. Table 3.2 outlines the types of SUT that could be used at each stage of this process, alongside the key physical attributes and process parameters to consider for each technology during the technical feasibility stage. In addition to parameters to be considered for a particular process step, material selection and the physical properties of the material (e.g., size limit, pressure limits) should be considered for all applications.

Once the options of an available SUT are narrowed down, the schedule of a trial should be considered and arranged. Most SUT suppliers offer systems that can be tested before use within the facility of the end-user, or the supplier. A trial should deliver preliminary evaluation of the suitability of the system and

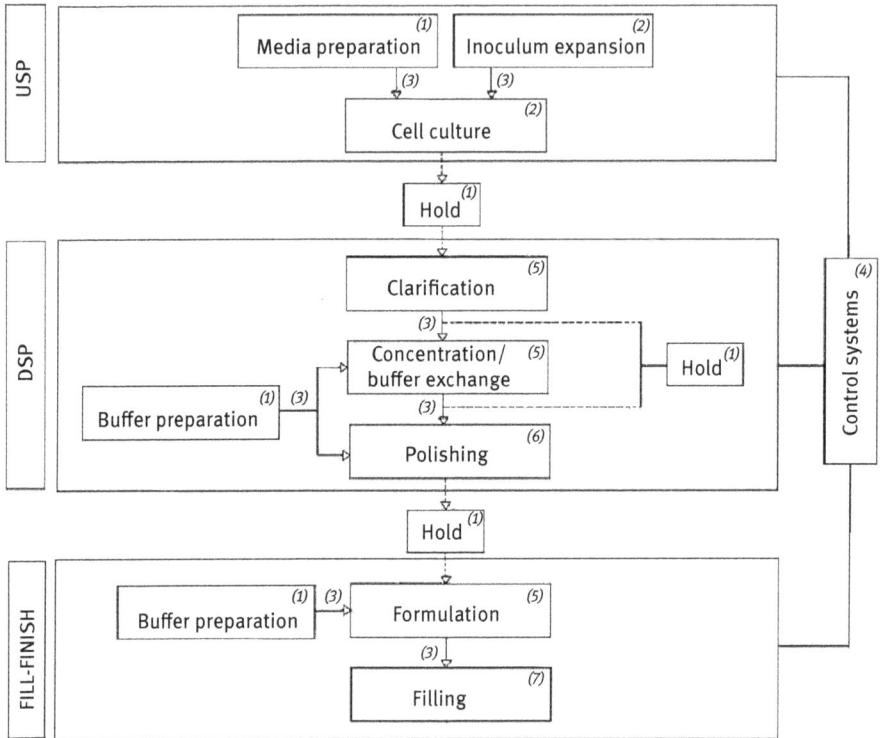

Figure 3.1: Typical operations in biopharmaceutical processes; (1) to (7) refer to SUT systems described in Table 3.2. DSP: Downstream processing; and USP: Upstream processing.

supplier, as well as start to build a relationship with the supplier. Single-use systems require an increased level of dependence on the supply of consumables, so building bridges with the potential new supplier from an early stage is advantageous. In addition to the technical trial, preliminary evaluation of the supplier's quality management system (QMS) should be undertaken. Some of the considerations to be taken during this stage are explained in more detail in Section 3.3. The SUT should be tested with the actual process materials and conditions or with a representative model system. By using the actual process conditions it is possible to undertake preliminary evaluation of the suitability of the design and system to control CPP and critical quality attributes (CQA) such as product stability or impurity removal. Choice of the team to undertake the trial is key to the success of the evaluation. Personnel with process (end-user's side) and system knowledge (supplier's side) are required to ensure success. A preliminary bill of materials must be prepared to undertake the trial. Therefore, an idea of how the system could be run, its suitability for processing, the complexity of the system, facility fit and logistics arrangements should also be considered early in the process. In parallel with the trial, more information can be

Table 3.2: SUT systems that can be used at each step of a biopharmaceutical process (Figure 3.1), and key physical attributes and process parameters to consider during evaluation of SUT feasibility [3].

Process flow	SUT systems	Physical attributes/process parameters
(1)	Pre-sterilised bags, liners and mixer- systems used for preparation, mixing, hold of intermediates as well as storage of media, buffers or final bulk drug substance, and waste collection	Integrity/leaks
		Sterility (if applicable)
		Leachables (particularly during storage of final bulk)
	Bag systems for storage and transportation	Temperature
		Handling
	Mixing systems: rocker, stirrer, recirculation loops	Mixing time, speed and homogeneity
(2)	Single-use systems used for cell culture consisting of a flexible bag cultivation container supported by a rigid (steel) housing that can be driven by mechanical (tipping, stirring, vibrating) or pneumatic (airlift, bubbles) means: rocker, SUB, hollow-fibre bioreactors	Sensors
		Integrity/leaks
		Leachables
		Sterility
		Transfer of mass and heat
		Growth performance
		Productivity
		Pressure limitation
(3)	Transfer lines: tubing, fittings, connectors, valves, clamps, pumps, sampling	Integration level/connectivity
		Integrity/leaks
	Connectors: sterile tubing, fittings or adaptors, and aseptic connection devices (welders, SIP connectors, sealers)	Sterility (if applicable)
		Pressure limitation
		Limitation of flow rate
(4)	Monitoring	Connectivity
	Sampling	Sterility (if applicable)
	Control systems (sensors)	Performance
	Sensors: DO, pH, conductivity, capacitance, flow meters, pressure, CO2, temperature. These can be pre- sterilised, re-usable or conventional sensors can be autoclaved and used	Robustness
		Calibration

Table 3.2 (continued)

Process flow	SUT systems	Physical attributes/process parameters
(5)	Centrifugation: single-use centrifugation cells and tubing lines for transfer of materials	Integrity/leaks
		Scaling
	Filtration: pre-sterilised filter capsules used for direct flow- filtration using a pad or encapsulated system applied for clarification, bioburden removal, sterilisation or viral reduction	Sterility (if applicable)
		Pressure limitations
		Limitations of flow rate
	Pre-sterilised filter cassettes or hollow-fibre modules for tangential- flow applied to microfiltration (clarification) and buffer exchange or product concentration	Mode of operation: normal flow *versus* tangential flow
		System flushing
	Fluid handling skid with single-use lines, sensors and control	
(6)	Chromatography: membrane adsorbers, monoliths, single-use pre- packed chromatography columns	Scaling
		Flow rate
	Fluid handling skid with single-use lines, sensors and control	Capacity
		Mode: flow through *versus* bind and elute
		Control and monitoring
(7)	Formulation	Sterility
	Sterile filtration	Leachables
	Final filling of vials or syringes	Particulates
	Pump systems: time–pressure peristaltic, diaphragm, rotary and piston-pump systems	Integrity/leaks
		Accuracy and precision
		Shear (pump, filling needle)

DO: Dissolved oxygen
SIP: Steam-in-place

provided by the SUT supplier to undertake more detailed component selection and system configuration design.

Assessment of the technical feasibility of the SUT is usually done after generation of a user requirement specification (URS), which outlines (in detail) the particular technical requirements for the equipment application. However, for SUT, it is recommended to start with a general list of requirements for initial discussions with suppliers, and adding greater detail when better understanding of the application is achieved. Knowledge will probably be gained from the feasibility stage which, once reviewed, can aid in writing of the detailed URS. This approach helps to reduce the resources required to be committed before the technical feasibility is established. Once the latter is completed, it is necessary to revise the URS to suitable levels of detail to ensure that the system meets requirements and can be validated accordingly. Further details on preparation of an URS can be found in Chapter 4. The URS should incorporate good engineering principles to the design and validation of the SUT, in the same manner as a traditional stainless-steel equipment/facility.

3.2 Business assessment

The feasibility of a SUT system should include a business case assessment to summarise the costs, benefits and risks of selecting a SUT for a given application. The assessment should be broad and consider the impact of the technology upon operations, the facility, supply chain and regulatory requirements. Then, these aspects can be coupled with the outcomes of the technical feasibility to identify whether the overall case for adoption of the technology is compelling and worth proceeding with.

Considerations for each of the wider business areas are summarised within Table 3.3. The first point to consider is to state clearly the scope for which the technologies are being considered. If the technology is to be applied to the manufacture of drug products, then any risks to product quality will probably have an impact upon patient safety, which in turn changes the level of acceptable risk. The level of experience of an organisation with SUT should be taken into consideration. If experience is high, then it builds confidence that benefits and risks can be identified. One outcome of the business assessment might be that the organisation needs to build experience with the SUT before the full scope can be implemented. This aim could be achieved by prototyping the technology or by partial implementation of SUT focussing on indirect product contact areas in the first instance. The main purpose of the assessment is to identify the benefits that might be realised by adopting the SUT, the costs associated with the change, and the risks to the business. Examples of risks and benefits for each of the main areas of consideration are also given in Table 3.3. The key benefits are often driven by simplification of operations, greater flexibility in facility utilisation, and reduction cost of goods sold (COGs).

Examples of the operational benefits that can be realised by utilising SUT are given in Figure 3.2. For the examples presented, use of SUT resulted in 54% reduction in total processing time derived from the reduction or elimination of CIP skids and SIP procedures. However, operational improvements vary depending upon application. For complex operations (in which a considerable number of tubing sets and bag assemblies require installing into the hardware), the set-up time might negate any savings from removal of CIP and SIP activities.

Table 3.3: Summary of business considerations, benefits and risks associated with implementation of SUT systems.

	Considerations	Benefits	Risks
Scope	Where within the process will single-use technologies be used: – Solution preparation and support activities – Generation of intermediate or bulk drug substance – Final drug product What experience does the organisation have with the technologies? Implementation cost	Improved COGs Reduced time to build process or facility to support product manufacture Flexibility to cope with changes in product demand	High cost of implementation Lack of knowledge in organisation to implement and operate SUT Impact product quality and risk to patient safety
Operational	Staff and resources required Personnel training Complexity of set-up, operation and disposal Health and safety Complexity of integration with hardware and control Robustness QC review *versus* consumable specification and release Product stability Contamination Batch changeover Product campaign changeover	Reduced reliance upon water and clean steam system Reduced risk of contamination due to failure in SIP or design flaws such as 'dead legs' Reduces risk of contamination across different product campaigns Shorter batch times and product campaign changeover	Process failure Complexity of assembly Interaction between consumable part and hardware Poor control/reproducibility Product stability Complexity of single-use assemblies Higher requirements for operator training Longer set-up time Risks associated with manual handling

Facility	Change in purpose (scale, product type, product number) Storage space (warehouse, in process) Footprint required Utilities requirement Requirements of room classification Generation and handling of waste	Flexibility to re-configure process rooms Removal of complexity of SIP and cleaning pipework Reduced capital cost through reduction in pipework/complexity of stainless steel Quicker build time Smaller footprint Easier to increase capacity to meet demand	Limitations on equipment size lead to an increase in the number of items within the facility and cleanrooms Increased warehouse space Staging areas required within processing rooms High flow of materials to be transferred into and out of processing cleanrooms
Regulatory	Product quality Impact upon existing manufacturing licenses Criticality of SUT: – Direct product contact processing technologies, product hold or product storage – Indirect product contact through preparation or hold of media, buffer solutions, cleaning solutions or water Validation of process and facility Maturity of technology	Removes SIP validation Reduces CIP validation	Supplier/materials qualification Reliance upon validation (e.g., sterilisation) by third party Validation of extractables and leachables Increased scrutiny from agencies if technology is immature and unproven

(continued)

Table 3.3 (continued)

	Considerations	Benefits	Risks
Supply chain	Number and cost of consumables or raw materials	Greater flexibility to meet changes in product demand (increase and decrease)	Cost of consumable part(s) increases significantly
	Quality systems (audit, QC specification and analysis) from the supplier		Greater reliance upon robustness of third-party systems for continuous supply of materials
	Robustness		
	Waste		
	Logistics (ordering and draw-down of materials)		Complexity of the supply chain increases
	Transportation		Quality failures of the third party

CIP: Cleaning-in-place
COGs: Cost of goods sold
QC: Quality control

Process activities:

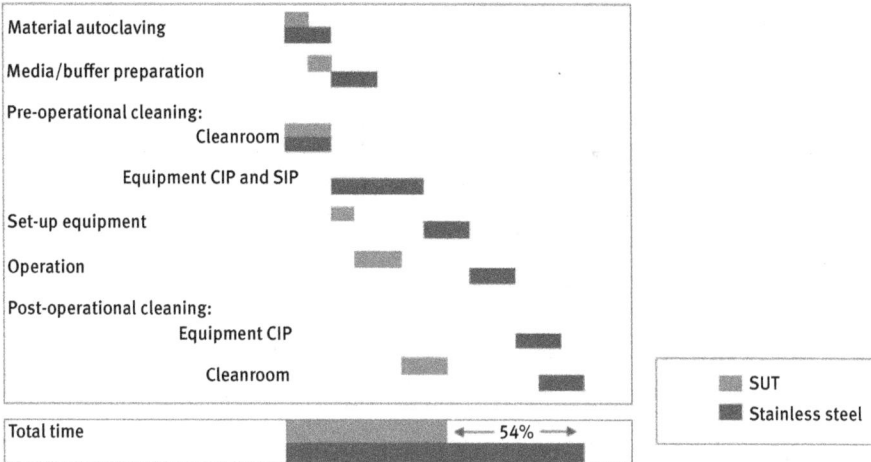

Figure 3.2: Process activities and timelines for stainless-steel equipment and SUT.

The use of SUT and their ability to make sterile connections has a major impact on the design and operation of the facility. The reduction in re-usable materials or components that require cleaning and sterilisation before use results in a reduction (or complete removal) of wash-up, cleaning and sterilisation areas. In addition, there is a reduction in utility requirements due to the removal of SIP and CIP flushes with water-for-injection (WFI) and/or purified water (PW), which in turn reduces the volume of liquid effluent. There is, however, a need for extended storage in warehouses for SUT components, incoming spares, and dealing with the increased quantities of solid plastic waste generated. Another impact is removal of media and preparation of buffers within processing rooms. This approach reduces the footprint of high classification areas, results in simplified workflows, and provides segregation between process operations and the preparation and handling of materials [4]. Increased manual handling as well as the health and safety considerations of this activity should be understood clearly when moving into SUT systems because larger quantities of liquids must be moved around the facility. Figure 3.3 presents a schematic drawing of differences between a typical stainless-steel facility and a facility using a SUT. SUT enables a closed-system operation that reduces the risk of cross-contamination, and protects products and operators. SUT minimise the impact on HVAC airflows and pressure differentials, and allows concurrent operations to take place in the same rooms. Preparation of tubing is undertaken in classified areas to ensure that standards of cleanliness are maintained, and that all tubing and connections are rinsed and assembled, and sterilised by autoclaving, or sent for external irradiation (if sterility is required). Once these materials are sterile, they are transferred to a hold area. A larger hall may be required to store and hold material before transfer into processing rooms.

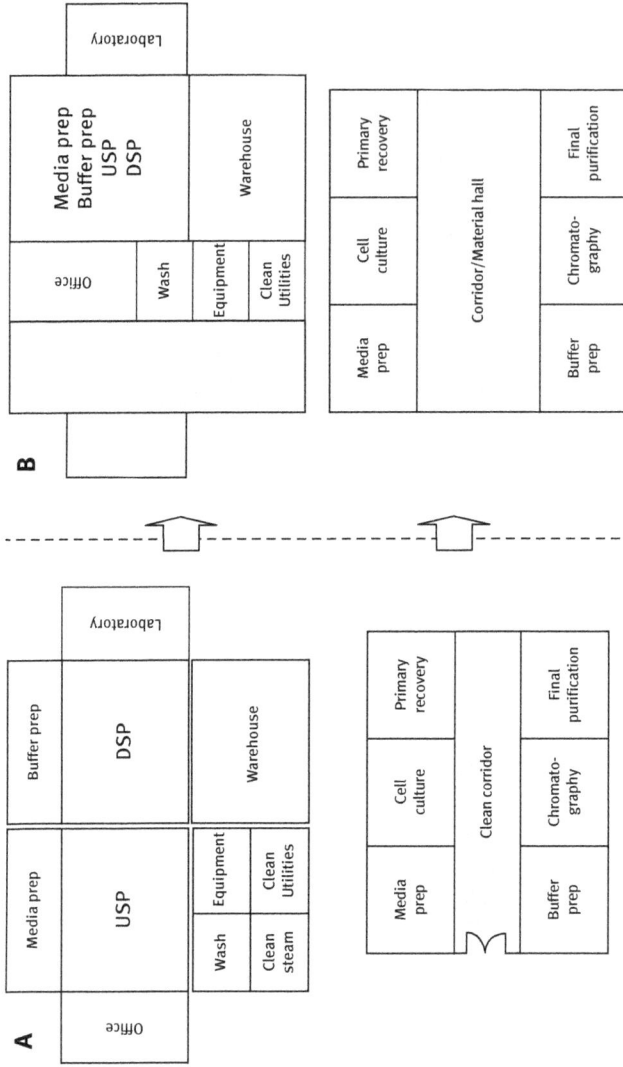

Figure 3.3: Facility schematics using (A) a traditional stainless-steel facility and (B) SUT facility.

The economic impact of the use of such SUT should be considered in terms of capital investment in the technology and wider facility, COGs, and the time required for implementation. Unless an organisation has direct experience with the technology to be implemented, it is likely that a modelling tool will be required to analyse the impact that the effect that technology might have upon COGs [5, 6]. Figure 3.4 presents, based upon a COGs model, an estimate of the percentage of savings when moving from a traditional stainless-steel facility to two scenarios that utilise different levels of SUT: (i) partial utilisation of SUT (hybrid), where all equipment is disposable with the exception of chromatography and filtration operations; and (ii) using single-use systems throughout the facility (disposable). In the examples shown, cost reductions were seen in cost areas such as capital, labour and utilities. Increases were seen in the cost of consumables and waste when compared with the stainless-steel facility. These findings correlate with what has been mentioned previously. That is, use of SUT systems reduces: capital costs due to reduction of the equipment/facility footprint; labour requirements, utilities reliance and water consumption (as a result of reduction or elimination of CIP and SIP operations); increase in generation of plastic/consumables waste.

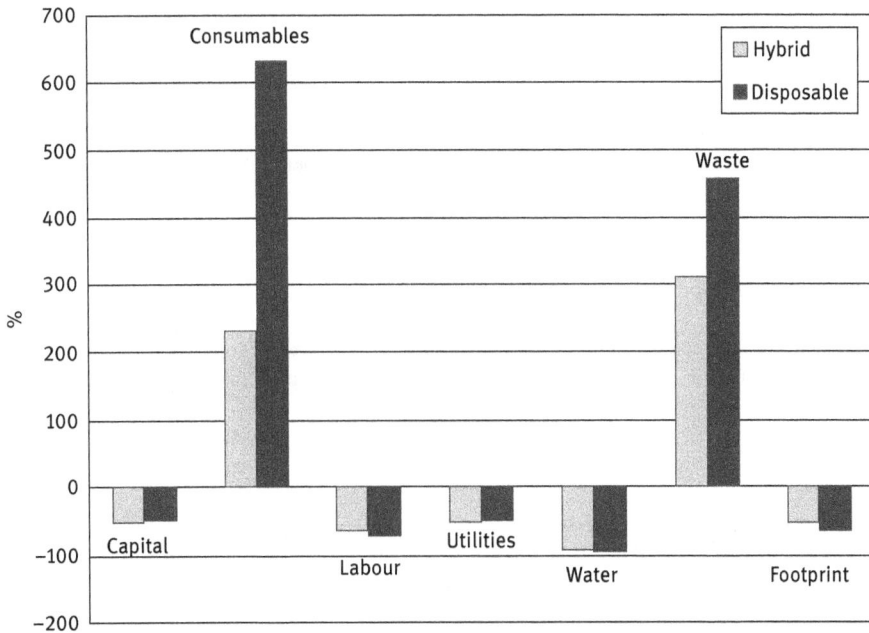

Figure 3.4: Estimated percentage of change in costs for a hybrid and fully disposable facility compared with a traditional stainless-steel facility [7].

Table 3.3 also highlights the many risks that may be realised if a SUT is selected and implemented incorrectly. Usually, the key risks are associated with maturity of the technology, supply-chain risks, and the level of experience that an organisation has with SUT. Moving to SUT means greater reliance upon the: (i) robustness of the manufacturing systems of the supplier for an uninterrupted supply of materials; and (ii) supplier QMS and their ability to deliver materials of the required quality for every batch. Selection of the correct supplier and qualification of supplied materials have a key role in mitigating these risks. Table 3.4 summarises a potential strength, weakness, opportunities and threats (SWOT) analysis for selection of a single-use supplier. This type of assessment is useful to summarise and present the positive and negative aspects of a given option. Weaknesses and threats are risks that may be mitigated as more experience is obtained with the supplier or technology, so the assessment should be linked to risks and updated throughout the project.

Table 3.4: SWOT analysis for selection of single-use supplier.

Strengths	Weaknesses
– Positive turnover and profitability.	– Highly bureaucratic
– Access to credit	– Limited product range
– Positive balance sheet	– Limited production capacity
– Established brand name	– Reduced inventory
– Spare product capacity	– Inadequate monitoring process or oversight
– Good location	controls of manufacturing process
– New manufacturing facility	– Unresponsive attitude to customer requirements
– Established customer base	– High staff turnover

Opportunities	Threats
– Trial of new technology	– Failure to meet quality requirements
– Technology development and	– Sources of materials or sub-assemblies
innovation	– Dependence upon single source providers for
– Sole provider of technology*	materials, assembly or irradiation could lead to
– Partnership for development of a new	interruption of supply (materials, assembly,
product range	irradiation and so on)
– Provision of technical expertise of a	– After sales service/technical support
specific product range (production,	– Continuity of supply of specific product range
testing, operation and so on)	

* May be considered as an opportunity for partnership (positive) or a potential threat to continuous supply (negative)

3.3 Selection of a supplier of single-use technology

Having a robust process to select, approve and manage a supplier ensures the quality and uninterrupted supply of SUT. The Food and Drug Administration (FDA), in conjunction with other regulatory bodies through the International Conference on Harmonisation (ICH), has outlined guidelines for implementation of a risk-based approach for selection of suppliers to minimise supply-chain risk, with increasing emphasis on controls around quality assurance of the product and security of supply.

ICH Q10 describes the pharmaceutical quality system, which extends to the control and review of the quality of purchased materials. It defines the end-user organisation as being responsible for ensuring that processes are put in place to ensure control of all purchased materials. It requires that processes incorporate quality-risk management, which includes the considerations set out below [8].
- Before selection of material suppliers:
    - Assess their suitability and competence to provide material using a defined supply chain by use of, for example, audits, material evaluation and qualification; and
    - Define responsibilities and the communication process for quality-related activities of the involved parties (e.g., written agreement between supplier and end-user).
- During use of materials:
    - Monitor and review the performance of the quality of the material supplied and the identification and implementation of any required improvements; and
    - Monitor incoming materials to ensure they are from approved sources using the agreed supply chain.

Management of risk starts with risk identification followed by a risk-mitigation strategy that requires planning, management and action. The end-user must consider risks associated with the SUT itself and the way it is manufactured and stored, as well as the potential supply and business risks that may affect supply. Table 3.5 lists risks that are associated with the supply of SUT (see also Table 3.4 for SWOT analysis of selection of a SUT supplier). These areas should be investigated by the end-user when assessing available SUT suppliers. It is unlikely that the full supply chain that leads to the fabrication of single-use materials will operate to GMP, so assessment of quality systems must be made to ensure that they are suitable for use in a GMP environment. The end-user should ensure that they have personnel familiar with the way that the SUT are fabricated so that they can assess whether the systems that the supplier has in place are suitable.

Table 3.5: Risk areas associated with supply of SUT alongside mitigation methods and sources of information that should be reviewed as part of the assessment process [9–11].

Risk area	Risk mitigation	Sources of information
	Suppliers – Product or process related:	
– Off-the-shelf/custom designed – Relevant tests (bursting, punch, seal tear and others) – Particulate contamination – Complexity and knowledge of manufacturing process – Sampling and QC testing	– Understand entire production process, definition of operating ranges and controls in place from raw materials to final assemblies – Manufacturing environment (cleanroom or ISO grade)	– Process description and mapping – Product quality review or annual product review * – Quality agreements with raw material suppliers
	Suppliers – Quality related:	
– Traceability of raw materials – Control and verification of manufacturing parameters – Lot-to-lot consistency – Maintenance and calibration program – Continuous improvement culture – Well documented manufacturing procedures	– Implement a quality agreement – Establish clear product specifications – Conduct robust supplier qualification process – Notification of changes of formulation components and/or manufacturing process	– Deviations/non- conformances – Near-miss events – Complaints and resolution – Internal/external audits – Investigations (people, premises, equipment, materials, quality assurance/QC, services, utilities and others) – Certificates of conformity

Suppliers – Supply management:

- Supply chain transparency
- Handling, storage and transportation fit for SUT characteristics
- Adequate supply management
- Service support

- Knowledge of complete supply chain mapping (organisations, controls, materials and services provided)
- Implement a supply agreement (estimated supply and costs)

- Raw materials and services supplier qualification program
- Change-control and notification from suppliers
- Data to support storage period and conditions of storage

Suppliers – Business-related:

- % supply/overall capacity
- Competing products that use the same production line
- Financial viability
- Disaster/contingency plan for supply

- Dual suppliers of raw materials
- Different manufacturing locations
- Available back-up equipment
- Availability of safety stocks of raw materials and assemblies

- Capacity increase/decrease versus capability
- Rate of company expansion/contraction
- Staff turn-over

End-users

- Having right skills in-house (quality, procurement, knowledge of how SUT materials are fabricated)
- Criticality and risk of SUT to the compliance of end-product
- Detectability of non-conformity of SUT supplied
- Single sourcing SUT?
- Supplier qualification: new or existing supplier?
- In-house storage space and conditions

- Develop a partnership with supplier
- Risk management process
- Implement key performance indicators (KPI)
- Identify second source of supply of SUT

* or review of the following documentation: completed batch manufacturing records, specification of starting materials, analytical data and in-processing batch testing for release of materials, Corrective Action and Preventive Action (CAPA), change-control, qualification of equipment, non-conformances and product recall, rejected batches and so on

ISO: International Organization for Standardization

QC: Quality control

The end-user is responsible for communicating and agreeing the product requirements with the SUT supplier. The end-user should request certification to ensure that:

- Adequate testing was done (Certificate of Conformity, USP class VI);
- Cross-contamination control was in place (i.e., dedicated equipment or validation of cleaning of non-dedicated equipment); and
- Full traceability of raw materials as well as certification that manufacture is free of transmissible spongiform encephalopathy (TSE)/bovine spongiform encephalopathy (BSE).

If SUT is product contacting, and if as a result of the specific risk assessment it is identified as high risk, the end-user may also request additional information. This information may concern the properties of plastic material used and/or results of any testing undertaken such as extractables profile, particle generation, and assurance that the material is free of chemical, microbial and particulate contamination (see Section 5.1 for specific details of materials qualification). The end-user should ensure that relevant personnel are involved in the specification, reviewing and evaluation of potential SUT to be used, and should include (as a minimum) a technical expert and quality representative (see Section 2.2 for details of teams and responsibilities, and Section 3.1 for considerations during feasibility assessment of SUT).

A risk assessment is an effective way to identify the controls required for uninterrupted supply of the SUT of required quality. Usually, SUT assemblies comprise various raw materials and components that are themselves sourced from different suppliers. Complete knowledge of the supply chain is important to manufacture the final single-use assembly and all organisations involved within it. Ensuring that a quality agreement is in place between the different suppliers and mapping all organisations involved in the final SUT assembly shows the security and authenticity of materials, and the services provided (e.g., sterilisation). The activities undertaken to manufacture the single-use component should be reviewed to identify those that are critical to the product and what could go wrong. An example of considerations for the selection and control of SUT products during supplier lifecycle management is shown in Figure 3.5. Processes should comply with relevant regulatory requirements (e.g., GMP) in addition to legislation for health, safety and environmental protection.

It is recommended to implement a robust supplier-qualification process that includes change-control and notification from the supplier of SUT of changes to the formulation or manufacturing process that may impact upon delivery of a consistent SUT product. Changes should be assessed based on risk assessment with impact based on criticality to the API. Reliability of supply is based on partnerships and agreements with suppliers as well as complete understanding and control of the SUT manufacturing process. A quality agreement between the supplier and end-user should establish clear responsibilities and clear product

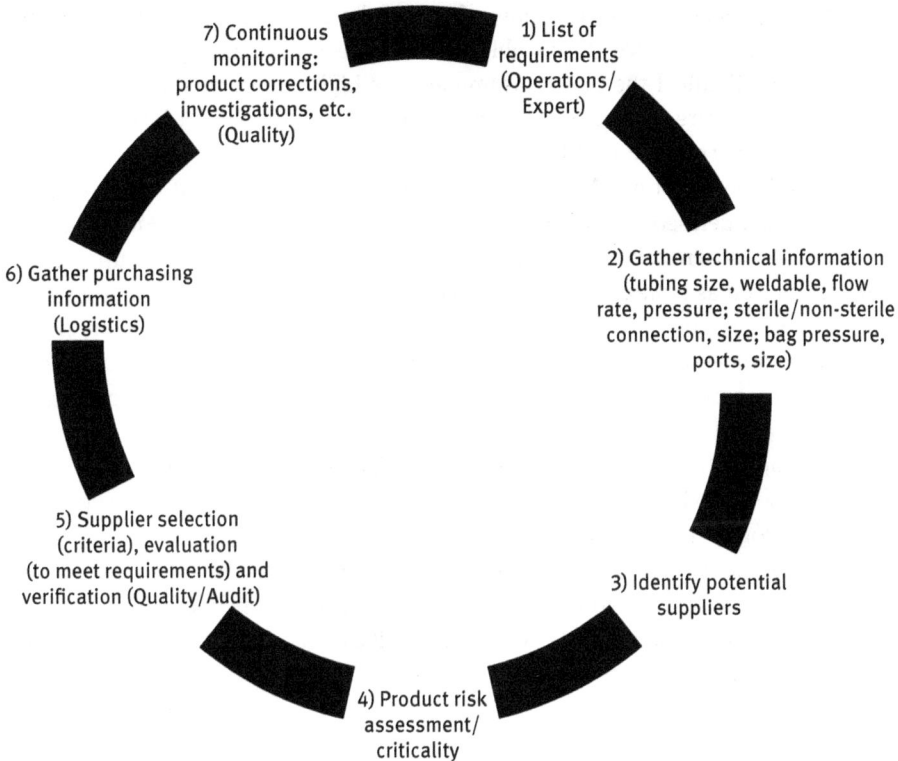

Figure 3.5: Guidance on selection and control of SUT products during lifecycle management.

specifications of the SUT. A consistent SUT product is ensured by establishing specifications, traceability and control of additives to the manufacturing process of plastic film. The supplier assessment must be an ongoing process that requires personnel with the appropriate skills capable of understanding the technical and quality profiles of the SUT manufacturing process to conduct a comprehensive audit of the supplier. Personnel capable of understanding the engineering aspects of SUT manufacturing and integrity of the supply chain are also required. As part of the continuous review of suppliers, the relationship with the supplier must be strengthened by implementation of metrics for performance indicators that can be tracked by both parties, as well as holding regular review meetings. Risk-management activities can form the basis for sharing identified hazards and mitigating risks. They can be used to demonstrate that all parties are taking a responsible approach to ensure the quality and safety of the SUT product as well as the security of supply.

For critical items, the end-user should aim to identify and implement a second source of SUT supply, and ensure that both SUT products are qualified in parallel.

This approach ensures that, if supply of any of the products is interrupted or discontinued, another source is available that will ensure continuity of supply. This aim may be difficult if the SUT consumables are highly integrated into a complex equipment hardware, but qualifying an alternative supplier will at least accelerate securing a supply chain if problems occur.

The end-user must ensure single-use materials can be received and stored, including having defined material specifications against which received materials are inspected. A larger number of materials required for a batch will impact upon warehousing, as well as storage within processing rooms. If a process uses a large number of single- use components, then having a staging area within the facility that is close to the processing rooms is advisable. This strategy allows for the materials required by the process to be batch-packed ready to be brought into the processing room, as required.

SUT suppliers will have the deepest understanding of their technology and its application. Their services can be used by the end-user if expertise and knowledge of these technologies in house is limited. Services provided may focus on: training of operators in the handling and operation of the SUT; development of written instructions and technical procedures; writing validation procedures and/or conducting validation activities such as installation and operational qualification (see Section 5.2 for more details on validation of SUT).

References

[1]   P.N. Hess and G. Dudziak, *Engineering Life Sciences*, 2014, **14**, 3, 332.
[2]   A.G. Lopes, T. Selas, A.G. Hitchcock and D.C. Smith, *Bioprocess International*, 2015, **13**, 9s, 1.
[3]   A.G. Lopes, *Food and Bioproducts Processing*, 2015, **93**, 98.
[4]   A. Sinclair and M. Monge, *BioProcess International*, 2004, **2**, 9, 26.
[5]   A.G. Lopes, A. Sinclair and N. Titchener-Hooker, *BioProcess International*, 2013, **11**, 8, 12.
[6]   A. Sinclair and M. Monge, *BioProcess International*, 2011, **9**, 9, 12.
[7]   A.G. Lopes in *Manufacturing Feasibility and Market Assessment of the Development of a Modular Manufacturing Facility for Local Production and Delivery of Inactivated Poliovirus in Low-and Middle-income Countries*, Part I Manufacturing Feasibility, LLB Global Health Solutions Ltd., London, UK, February 2012.
[8]   *International Conference on Harmonisation of Technical Requirements for Registration of Pharmaceuticals for Human Use, Pharmaceutical Quality System Q10*, ICH Harmonised Tripartite Guideline, Step 4 Version, International Conference on Harmonisation, Geneva, Switzerland, June 2008.
[9]   *A Guide to Supply Risk Management for Pharmaceutical and Medical Device Industries and their Suppliers*, Volume 1, Pharmaceutical Quality Group, The Chartered Quality Institute, London, UK, 2010, p.1.
[10]   C. Stock, *Contract Pharma*, January/February 2014.
[11]   J. Cappia, E. Vachette, C. Langlois, M. Barbaroux and H. Hackel, *Bioprocess International*, 2014, **12**, 5, 43.

4 Specifications and design of single-use technology

Once the scope of the application of the single-use technology (SUT) has been agreed and an initial business case has been approved, the next stage of the process is the specification, design and verification of the system. American Society for Testing and Materials (ASTM) International has published the E2500 international industry consensus standard for conducting a risk-based design of good manufacturing practice (GMP) systems. ASTM International recommends use of '*a risk-based and science-based approach to specification, design and verification of manufacturing systems and equipment that have the potential to affect product quality and patient safety*' [1]. This is a lifecycle approach that supports current regulatory guidance [Food and Drug Administration (FDA) and International Conference on Harmonisation (ICH)] based on risk, knowledge of the process and product, and quality standards.

ASTM E2500 emphasises that '*Good engineering practices should underpin and support specification, design and verification activities*' [1]. Good engineering practices (GEP) has been defined as the use of established engineering methods, standards and practices throughout the lifecycle of a facility to deliver cost-effective equipment/ systems fit for the intended use [2]. GEP is not mandatory by GMP regulations but is an effective implementation tool to drive technological innovation, cost efficiencies and compliance with regulatory initiatives. Considerations for GEP are that [1, 3]:

- Activities based on specification, design and installation should take full account of all applicable requirements (i.e., GxP, health and safety, environment, ergonomics, operations, maintenance, industry standards and statutory);
- Inclusion of adequate provisions related to quality into specification, design, procurement and other contractual documents;
- Production of lifecycle documentation covering planning, specification, design, criteria for verification, installation and acceptance, and service-maintenance requirements;
- Achievement of an appropriate degree of oversight and control through suitable verification of execution, construction and installation activities;
- Risk management to ensure balanced evaluation of risks against benefits, and identification of risk reduction to acceptable levels through design;
- Cost management to ensure that the cost impact of any activity is understood, assessed, and managed to return good value, which is measured as a balance between cost, quality and progress; and
- A clear structure of organisation and control is defined to respond effectively to the changing demands of the business.

https://doi.org/10.1515/9783110640588-004

GEP has been applied widely to stainless-steel projects, and is equally applicable to SUT. This is particularly the case if the SUT is new, or needs significant configuration to meet end-user requirements. A risk-based approach to the design process should be adopted. During SUT design, to set appropriate requirements and specifications the end-user must consider and incorporate relevant knowledge of the API and process from within the end-user organisation. The amount of knowledge available will vary depending upon the maturity of the product, the scale at which the process has been run and the level of experience that the end-user has with the SUT under consideration. If little knowledge is available in the first instance then a greater degree of testing will be required to build understanding as part of the design process and to reduce risk. If the scope of design project is broad, then greater design detail is required. Regulatory requirements and company-quality requirements will also need to be considered and included. This activity will tie in with the feasibility assessment and business case generated for adoption of the technology to generate a proposal for implementation and qualification of SUT. Knowledge and changes should be incorporated into the design project as more information is obtained, and reviewed periodically throughout implementation process. In this way, the risk profile of the design project will change and, ultimately, reduce as the project progresses.

4.1 Framework for the design project

A recommended framework for the design project is illustrated in Figure 4.1. The design process begins with a list of user requirements for each of the unit operations or systems for which SUT are under consideration. At this stage the requirements are not final or comprehensive, but they should have sufficient detail to define the intended use. They are not 'locked' at this stage because it is highly likely that they will change during the design process, and as the end-user increases knowledge (particularly if the SUT is new or the end-user has limited experience with the SUT). The list of requirements is sent to the supplier for review and for generation of a design with functionality that meets the requirements. Irrespective of whether an off-the-shelf or a custom-designed system is used, the end-user should verify the design as early in the design process as possible and before commitment of funds to the project to reduce risk that SUT is not fit for purpose. As a result, the end-user may require the need to trial the technology or test prototype before capital sanction is sought. The list of requirements and knowledge gained during previous steps will be converted into a formal user requirements specification (URS), in which the intended use and functionality of the system will be described in detail covering aspects related to the process, engineering, human interfaces, safety and the wider facility. Once capital funds have been released, the design of the system should be finalised and signed-off so that the system can be built, installed and qualified at the end-user site.

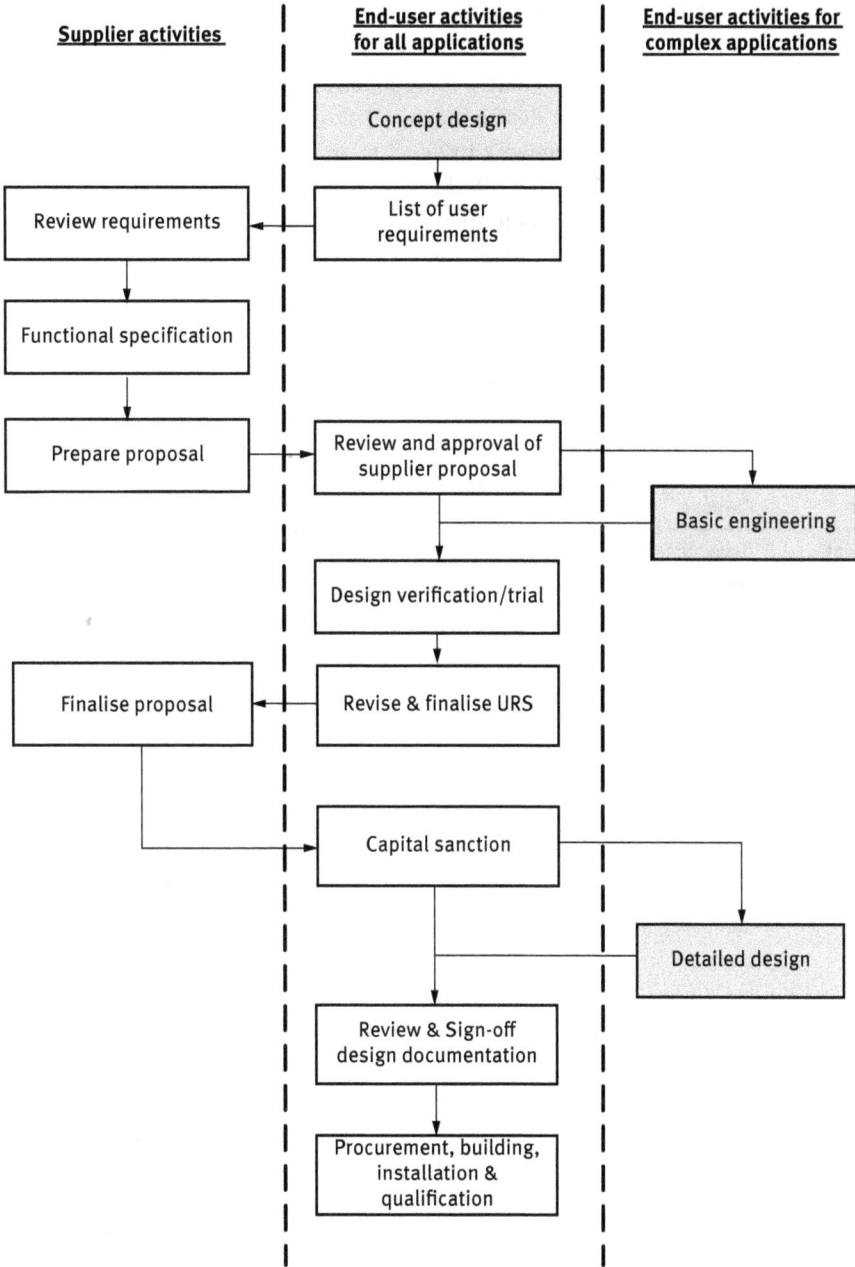

Figure 4.1: Framework for the design project of a SUT.

The framework for a design project presented in Figure 4.1 will vary depending upon the complexity of the design scope, with greater complexity requiring a greater number of formal design stages. Traditionally, complex engineering design projects go through the different design stages [2] shown in Table 4.1. These design stages provide a structured framework to develop and review how the SUT will be applied to a given process. Whether all of the steps are followed will depend upon the scope of application in addition to the maturity of the technology. For example, if a well-characterised and known technology is to be implemented, then only a concept study may be required. If a brand-new facility or major retrofit is required, then it is likely that a full design process will be required to ensure that a range of

Table 4.1: Stages of a complex engineering design project.

Stage	Purpose	Outcomes
Concept	Development of facility and system concept(s)	Architectural concepts of the facility
	Review requirements and design concepts	Requirements and constraints of the system
	Definition of new constraints and requirements	Process requirements
		Preliminary list of equipment
	Strategic planning for verification testing and project delivery	Estimate of scope and cost
		Milestone(s) schedule
Preliminary design/basic engineering	Development of system-level designs	Design P&IDs and drawings
	Definition of sub-systems	Lists of equipment and instruments
	Agree boundaries of the system and sub-system	Established management programme for engineering changes in scope, budget, schedule, and design intent
	Identification of components required to support system functionality	
	Complete strategic planning, develop execution-level planning	
Detailed design	Development and approval of component-level design specifications and engineering drawings	Installation drawings
		Details and specifications for procurement of systems and equipment
	Development of project schedule to support construction planning	Allowable methods and materials for fabrication and installation
	Purchase equipment with long lead times	
	Complete execution-level planning	Details and specifications for equipment acceptance

design approaches are considered before selection of the most suitable approach is made. Traditionally, the design approach described in Table 4.1 has been applied to facilities with a large degree of bespoke stainless-steel equipment that is fixed in place and connected to stainless-steel pipework. This approach is reasonable because design of this equipment must be developed to a high degree of detail before fabrication to reduce risk that the equipment will not meet end-user requirements. However, to achieve this level of detail requires the expertise of engineers. Typically, the end-user acquires this expertise by working in partnership with an engineering design company that will see the project through from concept to completion, and then handover to the end-user. The design approach requires significant investment in man hours, leading to cost and time to develop and refine the design. This cost is endured to reduce the risk of failure, ensure good engineering design practices and a robust process that can continue to support the product throughout its lifecycle. This approach is equally applicable to SUT, particularly for complex engineering design projects (e.g., if a new facility is being built or a SUT is to be implemented throughout the entire facility). As such, it will require a greater level of detailed design work that can be accomplished only by working in partnership with SUT suppliers and engineering companies.

4.2 Design choices and risk

For SUT, the end-user has the choice between an off-the-shelf integrated system or a custom-designed technology. Table 4.2 presents the design considerations and risks associated with these types of SUT systems. If obtained as an off-the-shelf item, the design of a final assembly is integrated into a support-and-control system developed and owned by the SUT supplier. Off-the-shelf systems have reduced customisation but the final assemblies are faster to fabricate and have reduced risk associated with start-up and qualification. Test controls can be re-used for several applications with only minor modifications. In addition, these systems have easier maintenance and reduced requirements for spare parts, which are added advantages. Choosing a supplier-designed system is also advantageous if the end-user does not have in-house capabilities to design a single-use system. However, once they have been fabricated and installed, there is a limited scope to make changes should the design not meet the requirements of the process. A supplier with a good track record of making single-use systems must be used. If the end-user does not have the skills to undertake or review the design, using an engineering design company or the design services of a SUT supplier may be needed to reduce risks. If a system is new to market, then the end-user must understand the degree of product testing that the system has been through before release, and review the results. The end-user may need to test the system to ensure that the design is robust, suitable

Table 4.2: Design risks for off-the-shelf and custom-designed SUT.

	Off-the-shelf SUT	Custom-designed SUT
Design ownership	Typically, the design is developed using supplier- provided components whereby the design is owned by the supplier. Very difficult to change supplier should problems arise with the design during verification (qualification) or production.	Custom-made configurations made by the end-user using 'standard' bags, and tubing sets/ connections. Design is owned by the end-user, so the design can be changed, as can the supplier of individual components, or dual- source suppliers can be used.
Design complexity	Systems appear simple due to removal of the requirement for CIP and SIP. However, complexity can be introduced rapidly through: – Interaction of single-use systems with equipment hardware and control elements; – Connectivity with stainless-steel equipment (hybrid system); – Transfer of process materials between unit operations; and – Large number of different single-use components used within the process.	
Design development	Supplier systems are often 'standard', so it is unlikely that they will be changed if problems arise during qualification. Ideally, potential issues must be identified before final selection of the supplier. Alternative options are: continue and hope to iron- out problems or switch suppliers; re-design with likely impact upon timeline. Available equipment cannot match the requirements of the design, leading to compromises. Custom tubing sets are often the best opportunity to generate standard bag designs.	Significant design efforts are required to develop the design of bags and tubing sets. Use of a custom-designed SUT requires the end-user to take on the responsibility for all design stages.
Maturity of the technology	Many single-use systems have been established for decades (e.g., bag systems, filtration). Other single-use systems are less mature (SUB for microbial applications, TFF system). Each technology should be assessed on a case-by-case basis: – How long has it been on the market? – How many design changes has it received? – How has the technology been applied to date (how many end-user clinical/com-mercial manufacturing sites?); and – What is the experience of the end-user with the system?	A custom-made system is likely to be immature if the design is specific to end-user requirements. Maturity can be developed within the design process by building prototypes and trialling to as large an extent as possible in an environment similar to the final application. Operators should also be included in the test/trial.

Table 4.2 (continued)

	Off-the-shelf SUT	Custom-designed SUT
Application	Options available across all scales from laboratory through pilot into clinical supply. Could be suitable for commercial supply, but dependent upon mass of product required and associated liquid volumes. More suitable for applications that require automation and control (providing that the supplier has built capability into the design).	Better suited for small-scale applications (laboratory or pilot). May not be suitable for clinical supply if a high degree of control and batch-to-batch consistency is required.
Manufacturability	SUT systems are effectively 'locked' in terms of capabilities. Due to a high level of operator interaction, the robustness of the equipment and ease of set-up should be an additional consideration. Application to the process necessitates assessment of how close to design limits the equipment will be operated, as well as the volume, pressure, and mixing. How well is performance of the single-use system defined and tested outside of laboratory test conditions? Scale-up from small-scale to large-scale manufacturing systems may not be straightforward.	Flexibility to tailor to the needs of the process, and assess robustness of design when used by operators. How well is performance of the single-use system defined and tested outside of laboratory test conditions? Practical performance of the technology may not be realised until day-to-day operations.
Operator interaction	Limited experience of the end- user with SUT: – No testing of equipment for final application before purchase. Ergonomics and usability of different configurations: – Manual long set-up times, failure due to incorrect set-up or due to too many components. New training requirements for set-up and testing of assemblies before and after use.	Scale of operation during process development different from scale technology been prototyped for final application. Depending upon the design and scale, limited or no automation can be available for a given application, leading to high interaction with operators during the process. Flow paths will require set-up and assembly to be in place before testing, operation and disassembly during and after processing. Scale and design could lead to ergonomic issues.

Table 4.2 (continued)

	Off-the-shelf SUT	Custom-designed SUT
Reliance on the supply chain	Manufacturing systems are highly dependent upon the supply chain to deliver all components (e.g., filters, resins, bag systems, tubing sets, container closures). Reliance on the supply chain increases with design complexity due to: – The variety of components available within the same category (bag, tubings, connectors) systems; and/or – Complex operations (e.g., TFF, chromatography).	Limited number of suppliers to choose from. If a small number of items are used in the manufacturing environment, then the cost per item will be high. Opportunities for dual supply if the end-user owns the SUT design.

CIP: Cleaning-in-place
SIP: Sterilisation-in-place
SUB: Single-use bioreactor
TFF: Tangential-flow filtration

for intended application, and is easy for operators to interact with, as well as ensuring quick installation of various single-use manifolds.

Alternatively, sometimes a custom-made design can be provided by the SUT supplier or other third party that can assemble the system from standard components, but this depends upon the technology application. In some instances, a combination of the two methods can be undertaken, whereby the off-the-shelf SUT assembly is changed to meet more specific requirements of the process of the end-user. A custom-designed system allows for greater control over the design of the system by the end-user. Developing custom assemblies can be advantageous because they can be tailored specifically to the requirements of the end-user, which leads to greater integration with existing operations, a simpler set-up, as well as reducing waste (e.g., optimised tubing lengths, filter positions, connections). In addition, items are delivered in their final assembly, which reduces the number of sterile connections made by the operator and reduces the risk of failure/contamination [4]. Irrespective of who is conducting the design of custom-made SUT systems, close collaboration with the SUT supplier is essential. Suppliers will help select the most suitable materials of construction, design and the tests to be undertaken in the final assembly to meet the specifications of the end-user. If more than one design option is considered, choices should be made based on cost, time, productivity and risk. At the end of the design stage, the supplier or third-party design provider will give a detailed drawing of the system, parts list and description of materials, as well as the support equipment/utilities required to operate the system. If the end-user is designing a custom-made SUT system then changes and improvements to the final assembly design can be made freely. However, the end-user may not have sufficient knowledge or experience in developing, designing and testing the SUT system.

The maturity of systems, their application and scale required should be considered during SUT design. Systems used for tangential flow filtration (TFF) are more complex and less mature than implementing bag systems for preparation and storage of media. Implementation of a single operation is less complex than retrofitting an entire facility (where different systems must be integrated together). Off-the-shelf systems supplied by different vendors have limited connectivity between different SUT systems. This problem can be overcome by working with the supplier or purchasing from a supplier that provides a complete offer of integrated or modular SUT systems that cover all process operations. However, these standard technologies may not always respond to the specific needs of the process in terms of feed rates, capacity, and automation. Also, reliance on a single vendor introduces supply-chain risk to the business, so a compromise in terms of process requirement instead may be needed. In terms of scale, custom-designed systems are more suitable for small laboratory- or pilot-scale applications because the systems are easier to assemble. However, control of the system is likely to require manual intervention by personnel during operation. This approach may be suitable for manufacture of pre-clinical or early clinical material, but is unlikely to be suitable for clinical or commercial supply, where batch-to-batch process and API product consistency is critical. In this case, a supplier-designed single-use system can be advantageous because suppliers often offer integrated monitoring/control systems that provide greater control of the process and fewer interactions with operators.

Other general considerations during the design of SUT systems are: a reduction in the number of parts; error-proof assembly design; ease of assembly; use of modular products. Product safety, ease of operator use, product yield and sterility assurance should be considered to minimise the risk associated with SUT assemblies.

4.3 Specification

Guidelines set by the European Medicines Agency (EMEA) state that '*The specification for equipment, facilities, utilities and systems should be defined in a user requirement specification and/or functional specification(s). The essential elements of quality need to be built in at this stage and any GMP risks mitigated to an acceptable level. The user requirement specification should be a point of reference throughout the validation lifecycle*' [5].

GMP regulations demand that the equipment be fit for intended use, so generation of a URS is essential irrespective of the nature of the equipment (i.e., stainless steel or single- use), if it is to be used in a regulated environment. The URS forms the basis of how the system will be validated during design verification (also known as 'design qualification') and during operational qualification. The URS should define technical, business and regulatory requirements clearly, and state precisely what the system should do, the functions to be carried out, the data on which the system will operate, and the operating environment. System requirements should be testable, measurable and traceable.

The URS is defined by the end-user, and the latter ensures that the technology is designed and delivered with the correct functionality. A URS should include the following general considerations:

- Key components and intended use;
- Process parameters: scale/volume, pressure, flow rate, process time, temperature, pH, and adsorption limit;
- Application requirements: integrity testing, sampling, process monitoring, integration with specific hardware, specific connections, utilities (compressed air, electricity, water with required purity level);
- Material requirements: cleanliness (USP Class VI, extractables/leachables, particles, pyrogens, free from animal-derived components) or microbiological levels (sterile or sanitised);
- GMP specific requirements: for example, system design in accordance with 21 CFR Part 11, and software designed in accordance with Good Automated Manufacturing Practice (GAMP ®) guidelines set by the International Society of Pharmaceutical Engineering (ISPE);
- Facility-related requirements: critical dimensions related to where the SUT will be situated or moved within the facility; environment that it will be exposed to (including cleaning agents); control-system interface with the wider facility system or other monitoring equipment; and
- Health and safety requirements.

If the system proposed to meet user requirements is complex and consists of multiple sub-systems, then design specifications are written to show greater detail of the system. In the example of a SUT there could be separate design specifications to cover the hardware, control system and single-use components that make up the overall, final SUT system. Examples of a URS for a single-use bioreactor and for TFF/diafiltration operations are presented in Chapter 6.

The next step of the process is to design a system to match user requirements. The design is captured within a functional specification (FS) which describes how the system will meet user requirements. In the case of SUT systems, the FS is usually defined by the supplier, but could also be developed in conjunction with the end-user if a customised design system is planned. The FS will specify materials of construction that meet requirements for operation (e.g., pressure, volume and mixing rates) and the level of cleanliness and sterility (e.g., provision of certification of sterility, USP Class VI and assurance that materials used for production are free of animal components). The FS will also state how any automated hardware used will meet requirements (e.g., 21 CFR Part 11 from the FDA or GAMP® guidelines from the

ISPE). Critical aspects resulting from URS and FS revisions as well as process and product knowledge should be identified and verified to mitigate high-risk aspects.

The technical subject matter experts (SME) of the end-user side will consider the operational compromises associated with a particular SUT system as part of a feasibility assessment, and will identify operational risks and define user requirements. Together with engineering personnel, they will identify the most suitable technology for a certain application. Then, the technical SME will determine how to test the SUT system and the work that needs to be carried out during verification and documented under GEP. The quality team will approve verification plans with other SME and ensure compliance with the quality management system on site. The Engineering team will inspect, test and validate the equipment/system against specifications. Manufacturing personnel will ensure operability of the system and generation of standard operating procedures.

4.4 Design verification

Design of a single-use system should be tested to verify that its function meets requirements. To facilitate this activity, prototype assemblies are designed and tested by end-users for fitness and function. The advantage of single-use systems is that, because there is no need to integrate them into facility support systems (e.g., steam), they can be tested externally, away from the final point of use. Hence, testing can take place in parallel without the need to access processing cleanrooms, which might be unavailable if they are being built, retrofitted or in use for production activities.

It is likely that the verification process will be an iterative cycle, with the system requiring a re-design after testing. The design-verification cycle is illustrated in Figure 4.2. The number of iterations will depend upon the complexity of the system that is being designed. In general, the more complex the system, the greater number of verification cycles that will be required. For example, a standard off-the-shelf design (which is not required to be integrated with other systems and is simple in its function) will not require as many re-design iterations as a bespoke custom-design system (which is highly connected into other systems and has complex operations).

Prototypes should be tested for normal operations and also under extreme conditions by conducting tests such as a final assembly integrity (leak) test. Other design considerations and testing are summarised in Figure 4.2. Compatibility of the materials that are in contact with process fluids should also be assessed. More details on this aspect of testing can be found in Chapter 5, which describes SUT materials and qualification of assemblies.

If a supplier is providing a custom-made design to meet functionality, any design modifications (e.g., deficiencies of the design, part substitutions, connections, tubing lengths, or support systems) must be agreed between the end-user and

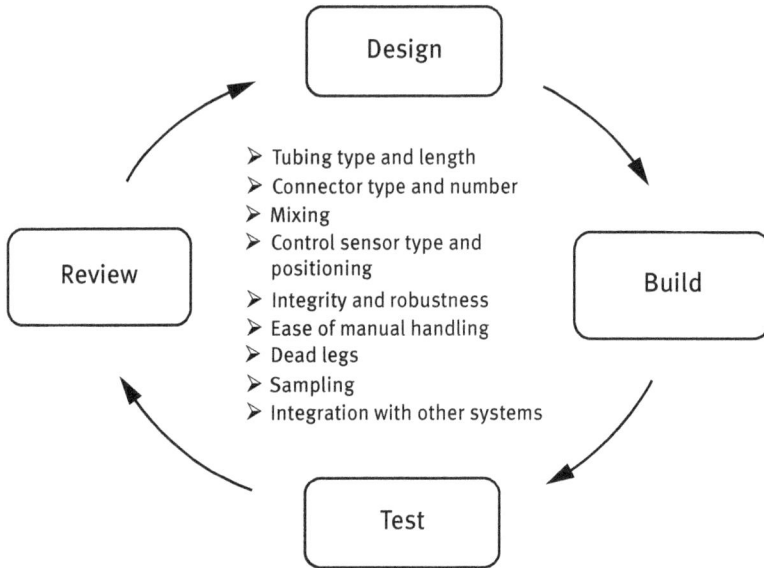

Figure 4.2: Verification cycle to test SUT prototypes and potential design considerations.

supplier. Initial testing can be undertaken by the supplier for design of the final custom assembly. For off-the- shelf systems, the end-user can schedule to trial the system to ensure that the design is suitable for the desired application. Many suppliers also have their own sites where a technology can be tested by the end-user. In the case of systems developed and designed solely by the end-user, the supplier of various SUT components must provide testing only for each of the individual components it supplies, and the end-user must ensure that correct testing is undertaken on the final assembly design.

'Verification' of the design differs from 'validation'. Verification identifies the best design to meet requirements. Validation demonstrates that the chosen design meets specifications consistently. The level of verification and validation should be based on risk, including those associated with product quality and patient safety, and the complexity of the system. Examples of specific design considerations and qualification of single-use systems used for storage and preparation of media, cell culture, tangential flow filtration, and filling applications are presented in Chapter 6. Validation of SUT systems are discussed further in Chapter 5.

References

[1]   *Standard Guide for Specification, Design, and Verification of Pharmaceutical and Biopharmaceutical Manufacturing Systems and Equipment*, ASTM International, West Conshohocken, PA, 2013.

[2]   *Science and Risk-based Approach for the Delivery of Facilities, Systems and Equipment*, International Society of Pharmaceutical Engineering, Tampa, FL, USA, 2011.

[3]   *ISPE Good Practice Guide: Good Engineering Practice*, International Society of Pharmaceutical Engineering, Tampa, FL, USA, 2008.

[4]   R. Wong, *BioProcess International*, 2004, **10**, 36.

[5]   *EU Guidelines for GMP for Medicinal Products for Human and Veterinary Use*, Volume 4, Annex 15, European Medicines Agency, Brussels, Belgium, 30th March 2015.

5 Validation

The Food and Drug Administration (FDA) defines process validation as *'the collection & evaluation of data, from the process design stage through commercial production, which establishes scientific evidence that process is capable of consistently delivering quality product'* [1]. This directive is in accordance with the International Conference on Harmonisation (ICH) guidelines for industry Q8 [2], Q9 [3] and Q10 [4], which underline a science- and risk-based decision-making approach within a quality management system (QMS) framework. This directive is also aligned with good manufacturing practice (GMP) regulations, whereby manufacturing processes must be designed and controlled to ensure that in-process materials and finished products meet pre-determined quality requirements, and do so consistently and reliably [5].

Validation can be a challenge if implementing single-use technology (SUT) as some of the methods differ from those used to validate stainless-steel equipment. For example, SUT does not require validation of cleaning and steam sterilisation that stainless-steel systems do. However, SUT require validation of the materials that enter into contact with processing stream, as well as proof that these materials do not affect the stability of the active pharmaceutical ingredient (API), or cause contaminants to leach into the process stream that cannot be subsequently removed. The approach for validating SUT is based on a combination of GMP principles and relevant parts of existing guidelines associated with a particular (medicinal) drug product. To ensure that validated materials, equipment and processes are used to manufacture the components that make up the single-use system, suppliers of SUT and end-users must work together to address specific SUT requirements. This effort provides the guarantee of reproducibility between batches.

At present, standardisation of SUT systems on the market is lacking. Differences arise from, among others, the material of construction (plastic material), components used, size, type of mixing, and sensor technology. These differences make validation activities highly dependent upon the type of technology selected. The other concern is to achieve a consistent supply of materials that are suitable for processing and that meet the required standard of quality. Disposable materials that are in contact with API product should be handled similarly to raw materials and as a part of the internal QMS. There is a lack of defined best practices for testing connectors, liners, tubing, and films, as well as no standard certification of materials from different suppliers. There is also no clear definition of what represents a 'qualified' SUT supplier. Verification and qualification of suppliers should include review of the QMS of the supplier to verify it meets standards such as the International Organization for Standardization (ISO) 9001 *'Quality systems – Model for quality assurance in design, development, production, installation, and servicing'*, ISO 13485 *'Quality systems – Medical devices – Supplementary requirements to ISO 9001'* or 21 CFR Part 820 *'Quality system regulations – Good manufacturing practices for medical devices'*. The most common SUT

https://doi.org/10.1515/9783110640588-005

issues raised during inspections of processes using SUT are related to equipment failures in processing, vendor qualification and/or criteria definitions for fitness-for-use, and documentation related to characterisation of extractable material [6].

Guidelines set by the European Medicines Agency (EMEA) state that only individual systems such as sterile filters and operations involving the final product (e.g., fill-finish operations or final bulk storage of drugs) must be tested [7]. In addition, there are no specific guidelines for other SUT used along the production process, and no studies concerning the impact that these technologies have on the final drug product [8]. Validation of SUT systems should fall within standard process validation activities whereby the end-user demonstrates process control. Material components and completed assemblies should follow a risk-based approach focused on impact upon critical quality attributes (CQA).

The decision whether to qualify a SUT system at the component level or at the application level depends on the risk assessment undertaken by the end-user. It is, however, recommended that a combined approach is followed whereby, at component-level qualification, test procedures are implemented for individual materials. Usually, these tests are undertaken by SUT suppliers if they have developed the design of the SUT system. The end-user should review all data provided by the supplier to ensure that the conclusions drawn are scientifically sound. This review is followed by a process qualification verifying that materials and final assemblies are fit for use and perform as expected for a particular application.

Assurance that the technical functionality of the SUT provides consistent and robust process control should be captured as part of standard approaches for process validation and exemplified by three process performance qualification (PQ) batches.

Based upon the most important issues raised, validation of the SUT can be differentiated into three key areas that must be maintained during supply of SUT products as well as during processing using SUT components:
1. Integrity of materials and final assemblies;
2. Materials used for assemblies are compatible with the API; and
3. Sterility and cleanliness of components are maintained while being used for processing.

5.1 Qualification of materials and assemblies

The first decision that must be made when designing and manufacturing a SUT system is the choice of materials to be used. SUT manufacturers must have their products tested thoroughly to comply with multiple qualification guidelines [9, 10]. Such testing includes qualification of raw materials, final assemblies and products (comprising physical, functional, chemical/biological tests) as well as sterilisation validation. Table 5.1 presents examples of the relevant guidelines for conducting tests that should be taken into account when qualifying materials or assemblies

Table 5.1: SUT materials certification and testing guidelines.

Category	Guidelines
Biocompatibility	ISO 10993-5:2009: Biological evaluation of medical devices – Part 5: Tests for *in vitro* cytotoxicity
	USP <87>: Biological reactivity test, *In Vitro*
	USP <88>: Biological reactivity test, *In Vivo*
Endotoxin	USP <85>: Bacterial endotoxins test
	EP 2.6.14: Bacterial endotoxins
Bioburden	In-house guideline based on ISO 11737-1:2006: Sterilization of medical devices. Microbiological methods. Determination of a population of microorganisms on products
Sterilisation	ISO 11137-2:2012: Sterilization of health care products – Radiation – Part 2: Establishing the sterilization dose
TOC	USP <643>: Total organic carbon
Conductivity	USP <645>: Water conductivity
pH	USP <791>: pH
Particulate matter	USP <788>: Particulate matter
Ions and oxidisable substances	USP <29>: Sterile water for injection
Extractables in model solvents	Extraction in a minimum volume (typically in water and ethanol)
	Quantitation by gravimetric non-volatile residue
	Qualification by Fouriert-transform infrared, high-performance liquid chromatography, or other applicable methods
	USP <1663>: Assessment of extractables associated with pharmaceutical packaging/delivery systems
Leachable testing	USP <1664>: Assessment of leachables associated with pharmaceutical packaging/delivery systems
Non-volatile residue	USP <661>: Containers – plastics
TSE/BSE animal component-free manufacture	EP 5.2.8
	Note for Guidance Revision 3 (EMA/410/01 rev. 3): Minimising the risk of transmitting animal spongiform encephalopathy agents *via* human and veterinary medicinal products

Table 5.1 (continued)

Category	Guidelines
Transportation	ASTM D4169: Standard practice for performance testing of shipping containers and systems
	ASTM D7386: Standard practice for performance testing of packages for single parcel delivery systems
	ISTA 2A Series: Partial simulation performance tests
Shelf-life	ASTM F1980: Standard guide for accelerated aging of sterile barrier systems for medical devices

ASTM: American Society for Testing and Materials
BSE: bovine spongiform encephalopathy
EP: European Pharmacopeia
ISTA: International Safe Transit Association
TOC: Total organic carbon
TSE: Transmissible spongiform encephalopathy

made of plastic. These tests are done under standard settings with standard solutions. Data from these tests should be made available by the SUT supplier to the end-user. The latter should audit the SUT supplier to evaluate the manufacturing process and testing of the SUT product to ensure lot-to-lot consistency.

The Bio-Process System Alliance (BPSA) is a trade association of the biopharmaceutical industry and supplies single-use components and systems for manufacturing processes. The BPSA has developed best-practice educational guides on quality tests for components [11], disposal [12], irradiation and sterilisation [13], and extractables and leachables [14–17]. These tests focus on individual single-use assembly components and final assemblies, packaging and transport (shipping), post-sterilisation testing and assurance of sterility, and integrity through product shelf-lives. In addition, SUT manufacturers must undertake qualification tests for plastic containers to ensure plastics do not change/degrade under conditions of normal use as well as plastic tests in accordance with the United States Pharmacopeia (USP) Class VI [14, 18–21]. Finally, SUT components and assemblies must prove they were not fabricated with animal- derived components and that they are free of transmissible spongiform encephalopathy (TSE)/bovine spongiform encephalopathy (BSE) components [22, 23].

As mentioned above, SUT validation focuses on integrity, compatibility, sterility and cleanliness. These criteria can be applied to qualification of materials and the final assemblies that must be maintained during supply of SUT. Verification should be undertaken for at least three different lots of materials supplied. Usually, this is the responsibility of the SUT supplier if the latter has developed the design of the final assembly. Qualification of material and assemblies should consider testing under extreme conditions to set the boundaries of operational limits.

5.1.1 Integrity

Plastic material used in SUT assemblies may be subjected to substantial mechanical forces during manufacturing, transport, installation and use. Therefore, the mechanical properties of raw materials used (hardness, tensile strength) must be considered carefully. Once SUT components are fabricated using the selected raw materials, they must be tested to show they do not rupture or are damaged in a way that makes them inoperable during processing. Rupture/damage can result in leaks, inaccurate readings from sensors, wearing of materials, failed filtration and potential process failures that can affect the quality of the API. Pre-treatment of plastic components (e.g., sterilisation) should also be considered because it can affect the properties of the plastic. As a general rule, specific process parameters resulting from validation studies and tests undertaken and defined by SUT suppliers should not be exceeded. The BPSA has outlined test matrices for plastic materials and assemblies [11].

5.1.2 Compatibiliy

Disposable technologies must prove that they are compatible with the drug product and that the latter is inert and does not react with the plastic material. According to GMP guidelines *'equipment shall be constructed so that surfaces that contact components, in-process materials, or drug products shall not be reactive, additive, or adsorptive so as to alter the safety, identity, strength, quality, or purity of the drug product beyond the official or other established requirements'* [24].

Tests to evaluate biocompatibility and to determine the release of extractable substances and particulates (see Table 5.1 for guidelines for testing SUT) must be done and collection of the resulting measurements and data should be made available from SUT manufacturer for individual materials. Standard methods for assessment of extractables [25] and leachables [26] in plastic containers have been established, and been applied to SUT. Recently, the Extractables Working Group of the BioPhorum Operations Group (BPOG) also made recommendations for standardised testing protocols for extractables in single-use systems during biomanufacturing [27].

5.1.3 Sterility and cleanliness

In cases for which bioburden control and sterility are required, the sterilisation method of SUT materials and assemblies must be validated to provide sterility assurance. Validation studies should be conducted to determine the efficacy of the sterilisation method. Specific load configurations, definition of dose range and temperature profile/locations of temperature sensors, biological indicators, type of materials, and compatibility with the sterilisation method should be documented in validation records. For autoclaves, this approach requires use of a defined load and operational cycle to check sterility after autoclaving. Annual checks with the validated load are required to ensure that the autoclave continues to maintain sterility. Once validated, the autoclave cycle should be maintained for manufacturing operations and records of the autoclave cycle attached to batch records. For gamma irradiation, a similar approach is taken whereby a dose-setting study is done to define the degree of irradiation required to achieve sterility before the dose is validated on a yearly basis. The dose level that the systems receive must be inspected with receipt of the single-use system before it can be released for use within manufacturing operations.

Testing of bioburden and endotoxins to obtain microbiological levels and total organic carbon (TOC) to assess levels of extractables, should be evaluated after sterilisation (see Table 5.1 for guidelines for testing SUT). The BPSA has outlined guidelines for irradiation and sterilisation of SUT materials and assemblies [13]. Gamma irradiation is the most common method used for polymeric material [28] and is the preferred option for SUT suppliers.

Typically, single-use suppliers provide the certificate of analysis for all tests undertaken regarding integrity, compatibility and sterility of the SUT component/assembly they supply. This certificate can be used by the end-user in place of carrying out the tests, but this should be evaluated in conjunction with the QMS qualification of the SUT supplier.

5.2 Process qualification

According to the regulations for GMP processing 'written procedures for production and process control which include sampling and testing of in-process materials and drug products need to be established to monitor outputs and validate the performance of process' [29], SUT suppliers are expected to provide end-users with information regarding testing for integrity/robustness, data (e.g., profiles of extractables and particulates) and (if required) sterility certification. Certifications such as the certificate of conformity, USP VI, TSE/BSE animal-free manufacture and sterilisation assurance should be provided for each lot supplied, and form part of lot-release information.

The end-user of SUT must demonstrate that selected materials and final assemblies will result in the target product being manufactured to a pre-defined quality under specified process conditions (process solution, temperature, time, pressure) and this is documented as part of the process validation. Additional studies and tests that need to be undertaken by the end-user should be evaluated by means of a risk-based approach using the qualification/validation documents of the SUT supplier [30]. These studies aim to determine the integrity, compatibility and (if required) sterility and cleanliness of the SUT component or assemblies with filled product under normal processing conditions.

The approach for qualifying a SUT system is not different from that used for traditional stainless-steel equipment (Figure 5.1). During process qualification the system design is evaluated to determine if it can deliver the functionality set out within the functional specification (FS) consistently, and ensure that it meets the requirements set out in the user requirements specification (URS). Qualification efforts should cover production equipment, including final assemblies, procedures and parameters. All process hardware, control devices, monitoring instruments and probes used for processing must undergo installation qualification (IQ), operational qualification (OQ) and, finally, PQ.

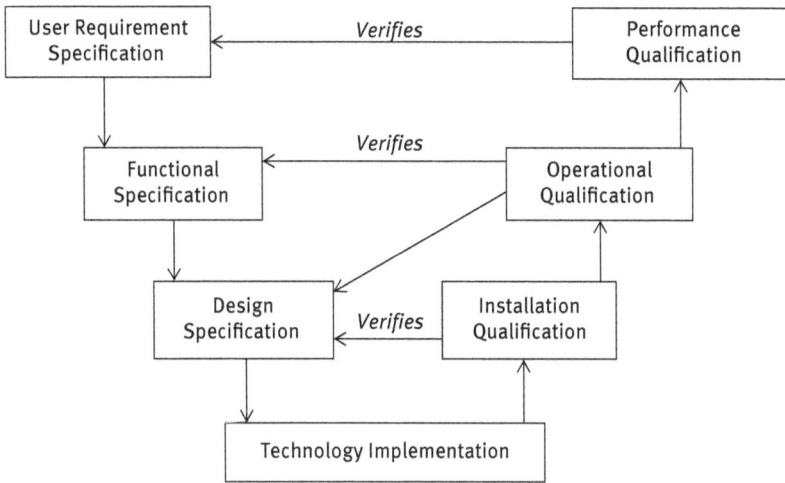

Figure 5.1: General framework for SUT validation divided into specification and qualification.

Collaboration between the end-user and SUT suppliers is crucial for implementation and validation of SUT systems. Suppliers must provide material specification with detailed assembly drawings, parts lists, and materials references. As an additional service, SUT suppliers can offer commissioning support with development of qualification protocols for site acceptance testing (SAT) and factory acceptance testing (FAT), as well as validation support with IQ and OQ services. During IQ/OQ, a validation expert from the SUT manufacturer together with end-user personnel ensure that equipment, hardware, instruments and disposable assemblies are designed and installed according to the design and FS. They undertake installation and set-up of the SUT system and provide on-site support for execution and implementation of qualification tests. Critical dimensions, design and part numbers of components, conformance to technical drawings and functionalities are documented in a pre-established and pre-approved IQ/OQ protocol. Verification that appropriate operating and maintenance documentation is available takes place, in addition to calibration of instruments and probes. Qualification of equipment and assemblies is completed with training of operators in charge of the system and establishment of documentation to be used during manufacture, such as batch instructions and standard operating procedures (SOP). In general, training can also be provided by the SUT supplier and must be documented as a part of GMP requirements. Chemical and biological tests can also be used to verify compliance with URS (if required). All changes brought about during IQ/OQ activities (e.g., re-design, assembly failures, technical investigations, performance/robustness improvements) should be controlled under change-control process. SUT suppliers should provide support for re-design/investigations engineering, maintenance and re-qualification of the system.

To carry out PQ of the final process, it is not sufficient to qualify only individual components. Final assemblies should be assessed under the same conditions as those described in the manufacturing instructions or SOP [i.e., with a suspension of the drug product or representative solution (assessed by risk assessment) and undertaken under normal production conditions (temperature, agitation, pH, pressure)]. PQ activity is the responsibility of the end-user and, in general, three consecutive PQ batches are run, which must show consistency in accordance with the process control strategy.

Depending upon SUT application within the process and the operational experience and process knowledge of the end-user, more than three PQ batches may be needed to validate the process. The earlier the end-user can gain experience with the SUT applied to the final process, the greater the chance that fewer PQ batches will be needed. In addition to specific considerations of the process, it is recommended that the PQ should also focus on demonstration of the three key criteria established previously: integrity, compatibility, sterility and cleanliness of the SUT systems used for processing. SUT suppliers should provide support to customers with regard to product compatibility and leachables through provision of data on extractables and development of protocols for particulates. If the SUT design was developed by the supplier, then any assembly failures or technical investigations resulting from use of the specific SUT system must involve the supplier who owns the final assembly design. Stated below are activities that can be incorporated into the Validation Master Plan (VMP) (some of which can be done concurrently with validation activities).

5.2.1 Installation qualification and operational qualification – Water or buffer runs

IQ and OQ activities can be used to prove that the SUT system can retain its integrity during processing. Disposable aseptic bags, tubing and connectors or final assemblies must be installed in a stainless-steel or plastic support for functional mixing, filling or transfer operations. Once assembled, the reliability of the single-use system can be validated. These activities are part of the IQ/OQ operations that are undertaken with a system filled with water or buffer. The SUT system is evaluated before and after processing under normal operating conditions and as a part of a IQ/OQ protocol. Pre- and post-integrity tests may be required to detect leaks and perforations in the system that may have occurred during operations. Depending on their applicability, SUT modules are checked visually for integrity and leaks using physical and mechanical tests (e.g., module integrity, leak test). Post-operation SUT modules can be tested further with destructive or non-destructive tests (e.g., integrity and burst strength tests). For example, filters are usually tested pre- and post-operations to ensure they have not been damaged. If a sterilising-grade filter is employed, the integrity result is used to prove that sterility capacity was maintained during processing. Pre- and post-operation leak tests may be required for SUT systems that must be sterile during operation or during storage. Specific tests and guidelines have been outlined by the BPSA [11].

5.2.2 Process simulation – Media fills

Process simulations or media fills involve validation of aseptic operations (e.g., fill-finish assemblies and operations) or aseptic connections and transfer activities. They necessitate exposing all product contact surfaces (tubing, connectors, containers) of the system to be tested with microbiological growth media (e.g., tryptic soy broth) under an identical environment, process manipulations and process conditions to simulate closely the same exposure that the drug product itself would undergo. Then, liquid media are incubated and assayed for bacterial content to demonstrate the capability of the process to produce a sterile product reliably and repeatedly. The FDA provides comprehensive guidance on validation of sterile processing activities that can be used for any system [31].

5.2.3 Performance qualification – API stream

Compatibility of a SUT system with product and process parameters should be tested. Tests to determine and quantify released leachable substances and particulates generated during processing must be done. Samples from the SUT system can be collected over time to assess for leachables and any particles that may be produced during processing (derived from filters, pumps and valves). A 'model' buffer or water can be used to obtain levels of extractables and particulates in the first instance, but the interaction of these substances with a specific API stream may differ from the model fluid. As such, an explicit profile of leachables and particulates must be tested and presented for each API, and tested under the same conditions as described in the SOP. Samples must be taken over time to quantify the levels of these substances during processing. If the API is highly valuable and/or a larger-scale of operation must be validated, a small-scale system that mimics the final assembly can be used to undertake preliminary quantification and assessment of the impact of these substances upon the API. Examples of how to correlate the results obtained from this study with large-scale operation can be obtained from Section 6.1, which relates to the data analysis of an extractable profile for bag systems. Other examples of specific requirements for qualification of process equipment (e.g., single-use bioreactors, single-use equipment for tangential-flow filtration, fill-finish operations) are shown in Chapter 6.

5.3 Continuous improvement of processes

According to the FDA, improvement of processes can be achieved by innovation, continuous improvement and as a result of corrective actions [32]. This is a combined result of: audits; analyses of batch data; corrective, preventive and improvement actions; and management review. Innovation is a consequence of a change of direction and investment in alternative technology or systems by the company and involves technical experts. Efforts in continuous improvement are a result of process optimisation (ICH Q8) and a higher level of process understanding. Corrective actions are needed if out-of-specification, procedural deviations or investigations occur.

Continuous improvement is essential in a QMS. As stated by the FDA, it '... *aims at improving efficiency by optimising a process and eliminating wasted efforts in production...and reducing variability in process and product quality...*' [32].

Continuous verification of processes focusses predominantly upon review of the process-control strategy to assess if modifications are required as more knowledge of the process is obtained, due to the increasing numbers of batches being completed. Carrying out the SUT may impact upon overall control of the process, but other factors should be monitored in relation to SUT. These concern the ease of use and set-up when operators work with such systems, as well as the consistency and resulting quality of supply of SUT. When trending and reviewing the interaction of operators with the SUT, it is important to identify any deviations or batch failures that could arise due to the complexity of handling SUT or the incorrect set-up of the systems. Additional training might be sufficient to address these issues, but they might also arise due to design flaws in the SUT systems that become apparent as more batches are run. For SUT designed and supplied by third parties, solutions to address these issues rapidly may be difficult to find, so developing and implementing alternative options to the SUT may become necessary. This approach is likely to be inefficient, so it is preferable to avoid this situation by identifying issues during earlier design-verification tests and/ or trial runs, when there is more time and opportunity to make changes.

The quality and robustness of the SUT supply chain should also be reviewed to ascertain if supply risks could impact upon manufacturing operations. Supplier assessment is an ongoing process throughout the SUT product lifecycle, and performance data should be reviewed directly with the supplier. If supply issues are experienced, it might be necessary to identify an alternative supplier/technology that can be validated and used as a back-up in case unacceptable issues arise or ongoing problems are not resolved. The level of qualification of a SUT system is based on criticality assessment of the SUT to the API. Process changes communicated *via* a change notification should be assessed based on risk. For example, if the SUT is product contacting, changes in compositions of plastic material may result in re-qualification of the material to ensure a similar profile of leachables. Changes in design that may be a result of changes in internal processes (e.g., changes from using weldable tubing to sterile connections, which could be a change implemented by the supplier) should also be assessed. If the SUT design is owned by the end-user, changes will be controlled under a change-control system and could involve an alternative supplier. If the design is owned by a third-party supplier then, unless the SUT has been adopted widely and the issues are seen across many end-users, influencing the third party to re-design the system could be difficult. As such, it may be necessary to evaluate alternative technologies again with a view to their implementation within the process. If changes in design are driven by a change in the SUT portfolio, the end-user must assess the suitability of the new design to its needs.

If using SUT, the number of components, systems and assemblies should be reduced (if possible) into a smaller portfolio because it results in simpler management of stock, greater leverage over suppliers and improved change management. If the SUT is an off-the-shelf system, the SUT supplier should provide support maintenance and re-qualification services to support equipment.

Issues may arise only when the SUT has been exposed to a large number of batches. However, this is the point at which it is most costly and painful to make changes. Therefore, the greater the number of issues that can be identified early in the SUT design stage, the lower the risk to the end-user of failures having a significant impact upon operations.

References

[1]   *Guidance for Industry Process Validation: General Principles and Practices*, Revision 1, Department of Health and Human Services, Food and Drug Administration, Silver Spring, MD, USA, January 2011.
[2]   *ICH Harmonised Tripartite Guideline: Pharmaceutical Development Q8(R2)*, Step 4 Version, International Conference on Harmonisation of Technical Requirements for Registration of Pharmaceuticals for Human Use, Geneva, Switzerland, August 2009.
[3]   *ICH Harmonised Tripartite Guideline: Quality Risk Management Q9*, Step 4 Version, International Conference on Harmonisation of Technical Requirements for Registration of Pharmaceuticals for Human Use, Geneva, Switzerland, November 2005.

64 — 5 Validation

[4] *ICH Harmonised Tripartite Guideline: Pharmaceutical Quality System Q10*, Step 4 Version, International Conference on Harmonisation of Technical Requirements for Registration of Pharmaceuticals for Human Use, Geneva, Switzerland, June 2008.
[5] *Pharmaceutical cGMPs for the 21ˢᵗ Century – A Risk-based Approach*, Department of Health and Human Services, Food and Drug Administration, Rockville, MD, USA, September 2004.
[6] A.M. Trotter in *American Pharmaceutical Review*, American Pharmaceutical Review Fishers, IN, USA, 30ᵗʰ March 2012.
[7] *Guideline on Plastic Immediate Packaging*, EMEA/CVMP/205/04, Materials Committee for Medicinal Products for Human Use, Committee for Medicinal Products for Veterinary Use, European Medicines Agency, London, UK,19th May 2011.
[8] A.G. Lopes, *Food and Bio-products Processing*, 2015, **93**, 98.
[9] J. Martin, *Bio-Process International*, 2007, **4–5**, 1.
[10] D. Eibl, R. Eibl and P. Kohler, *DECHEMA Biotechnology*, 2002, **2**, 1.
[11] Bio-Process Systems Alliance Guidelines and Standards Committee, *Bio-Process International*, 2007, **5**, 4, 52.
[12] Disposals Subcommittee of the Bio-Process Systems Alliance, *Bio-Process International*, 2007, **5**, 22.
[13] The Irradiation and Sterilization Subcommittee of the Bio-Process Systems Alliance, *Bio-Process International*, 2007, **5**, 10, 60.
[14] Bio-Process Systems Alliance Guidelines and Standards Committee, *Bio-Process International*, 2007, **5**, 11, 36.
[15] Bio-Process Systems Alliance Guidelines and Standards Committee, *Bio-Process International*, 2008, **6**, 11, 44.
[16] *Millipore Technical Brief*, Millipore, Billerica, MA, USA, 2010.
[17] B. Caine in *Bio-Process Systems Alliance*, Washington, DC, USA, 2009, pp.1–27.
[18] V. Anicetti, *Bio-Process International*, 2009, 7, S1, 4.
[19] E. Jenness and V. Gupta, *Bio-Process International*, 2011, **9**, S2, 22.
[20] D. Bestwick, R. Colton, Bio-Process International, 2009, 7, S1, 88.
[21] J. Furey, Millipore Corporation, Billerica, MA, USA, 4ᵗʰ May 2004, pp.1–30. [Private Communication].
[22] EP 5.2.8, 01/2008:50208, European Pharmacopeia, Strasbourg, France, p.539.
[23] *Official Journal of the European Union*, EMA/410/01 rev.3, 1–18, European Commission, Brussels, Belgium, 2011.
[24] *GMP Requirements*, Department of Health and Human Services, Food and Drug Administration, Rockville, MD, USA, April 2007, Volume 4, Part 211.65(a), Revised 1.
[25] USP <1663>, First Supplement to USP 38-NF 33, 7166–7180.
[26] USP <1664>, First Supplement to USP 38-NF 33, 7181–7193.
[27] W. Ding, G. Madsen, E. Mahajan, S. O'Connor and K. Hong, *Pharmaceutical Engineering*, 2014, **34**, 6, 74.
[28] M.R. Sagheem and T. Sanle, *Pharmaceutical Technology*, 2012, **36**, 5.
[29] *Pharmaceutical cGMPs for the 21ˢᵗ Century – A Risk-Based Approach*, Department of Health and Human Services, Food and Drug Administration, Rockville, MD, USA, September 2004, pp.1–32.
[30] B.I. Barnoon and B. Bader, *BioPharm International Supplements*, 2ⁿᵈ November 2008.
[31] *Guidance for Industry Sterile Drug Products Produced by Aseptic Processing – Current Good Manufacturing Practice*, Department of Health and Human Services, Food and Drug Administration, Rockville, MD, USA, September 2004.
[32] *Innovation and Continuous Improvement in Pharmaceutical Manufacturing: Pharmaceutical cGMPs for the 21ˢᵗ Century*, PAT Team and Manufacturing Science Working Group Report, Food and Drug Administration, Rockville, MD, USA, 2004.

6 Case studies

Numerous unit operations in a biopharmaceutical production process can take advantage of single-use technology (SUT), from bioreactors, filtration, centrifugation chromatography and filling operations through to mixing, connection/disconnection, sampling, product transfer, hold and storage. Even though one unit operation may involve a single application, several types of SUT (e.g., bags, filters, tubing, tubing, fittings, sensors, valves) may be involved, all of which are composed of materials and properties that may differ.

Implementation of a SUT system starts with selection of the appropriate technology for the intended use, followed by a detailed definition of the intended application of the equipment and all related requirements. This information should be captured in a user requirement specification (URS). Based on the URS, the design phase and material selection of the SUT is initiated. System design of the SUT will be draft versions until a feasibility trial or verification is conducted before selection. During this stage, relevant component- and product-based tests can be established. At this point, understanding of the requirements and preferred options should be sufficient to reassess capital sanctions.

Once capital approval is assessed to acquire the technology, then the URS, SUT design and material selection are revisited and completed. The final stage is qualification of the process equipment parameters and procedures, followed by a final process performance qualification (PQ). This chapter presents a general overview and examples of particular attributes and considerations during the selection, specification, design and qualification of SUT used for:
- Storage and preparation of solutions using bag systems;
- Cell culture using single-use bioreactor (SUB) technology;
- Tangential-flow filtration (TFF); and
- Formulation, sterile filtration and filling operations.

Considerations during implementation of off-the-shelf and custom-designed SUT systems are presented.

A risk-based approach is presented throughout case studies, with examples of more formalised versions of risk assessments, such as failure mode and effect analysis (FMEA) applied to design, operation and quality attributes of the product.

All examples shown are for demonstration only because each organisation will have their own methodology and approach to assess risk.

All sections within the case-study chapter must be read to elicit good understanding of overall considerations of SUT implementation because different examples are presented in different sections, which can be applied readily to a given SUT system.

https://doi.org/10.1515/9783110640588-006

6.1 Case study 1: Single-use bag systems

Single-use bags are very well-established technologies that have been used for >20 years within the biopharmaceutical industry. They can be used for processing operations and support activities. Their applications are essential for aspects of bioprocessing and include:
- Preparation of media and buffers;
- Hold and storage of media and buffers;
- Preparation and hold of cleaning solutions;
- Intermediate process hold;
- Storage of final bulk drug substance;
- Collection of samples and waste;
- Cell culture and microbial fermentation;
- Mixing of process intermediates;
- Freezing and thawing of bulk drug substance or intermediates; and
- Shipping of product, intermediates or solutions between manufacturing sites.

Single-use bags are available in various sizes, shapes, configurations and different types of plastic film. They are available from open liners to closed systems such as pillow-type two-dimensional (2D) bags (1–50 l) and box- or cylindrical-shaped three-dimensional (3D) bags (50–2,000 l). All systems require support of an outer support container that can be a flat platform (2D bags), or cylindrical or rectangular in shape (3D bags). By using single-use bag systems, the end-user benefits from closed processing, reduced risk of contamination and simple reconfiguration of manufacturing operations. Greater flexibility on volumes and configurations is provided by the wide range of bag types and sizes that can be increased or decreased according to production needs.

Bag systems require minimal pre- and post-use activities, with no cleaning [clean-in-place (CIP) and steam-in-place (SIP)], leading to low impact upon secondary-support infrastructures and equipment in the facility. Bag systems can reduce the logistical complexity by use of standard designs, types and sizes with pre-connected tubing and connectors. Despite being an established technology, single-use bags have limitations based on: size and technical operation; a higher degree of manual and operational handling; and risk of potentially toxic or inhibitory substances (leachables) from plastic materials entering the API product stream.

6.1.1 Material selection

Typically, bags are made of multi-layer thermoplastic films that are irradiated. Multi-layered films consist of:

1. An external structural layer [e.g., polyamide (PA), polyethylene terephthalate (PET), low-density polyethylene (LDPE)] with heat-resistant and mechanical properties;
2. A middle layer which provides a gas barrier and water-vapour properties [e.g., ethylene vinyl alcohol (EVOH), polyvinylidene chloride (PVDC), PA]; and
3. A fluid-contacting inner layer [e.g., polyethylene (PE), ultra low-density polyethylene (ULDPE), ethylene vinyl acetate (EVA)] that should be inert and have good sealing properties.

Each film is composed of a mixture of homologous polymers and other substances such as plasticisers, stabilisers, antioxidants, pigments and lubricants. These additives provide the plastic films different physical and protective properties (e.g., flexibility, rigidity, stability barrier). These additives are also a major source of leachables, which could migrate from plastic into the solution contained within the bag. Leachables present a risk of contamination to the product that could impact upon its structure and activity, or pose a toxic risk to patients. Besides the composition and processing method of the plastic, cleaning procedures, surface treatment, contact media, inks, adhesives, permeability of preservatives, sterilisation conditions and storage conditions also promote leachables. These aspects should be considered when assessing the suitability of a particular single-use container to a given application. For example, EVA has low moisture and gas-barrier properties and requires a secondary barrier. However, it does not contain plasticisers, which reduces the amount of leachables, and hence is a good film for the inner layer. EVOH has high gas-barrier properties and moderate water vapour [1, 2], and so is a good film for the middle layer.

The first step in selection and evaluation of the appropriate bag system concerns the suitability of the film material for a certain application. First, the plastic material must be suitable for the solution it will hold. That is, it does not interact or change the solution, and it provides the correct containment level and barrier properties. Loss or entry of water, gases or solvent may occur through the permeable container surface. SUT suppliers should provide test results conducted for biocompatibility, cleanliness (endotoxins, particulate level, sterility) and an extractable profile of films (though the end-user may be required to purchase this information). Table 5.1 presented in Section 5.1 shows the required testing and certification of SUT materials that should be obtained from the supplier, and which is considered to be the minimum requirement for an initial assessment of suitability for single-use bag systems.

The second consideration is the integrity of the plastic material and whether it remains robust under normal operating conditions. Culture bags may contain internal moving parts (agitation system), carry large volumes, and be submitted to large pressures and variable temperatures. Tubing may be used inside a peristaltic pump head in which repeated compressions of the tubing wall degrade the material, thereby increasing the potential for leaks. Mechanical properties such as hardness, tensile strength and modulus, melt flow, as well as testing of seals, bursts, drops, air leaks, ships and microbial ingress help determine the critical and breaking point of these materials. The Bio-Process System Alliance (BPSA) has provided a comprehensive guide for testing of individual components such as bags, tubing, connectors and filters [3]. The SUT supplier should be able to provide certification of tests done to each SUT component or to the final assembly.

6.1.2 Risk assessment for extractables and leachables

Specific standards on how to assess extractables and leachables from single-use bioprocessing materials are lacking. However, there are pharmacopeial and industry guidelines that can be applied to these systems: final containers and closures for orally inhaled nasal products [4] (where the risk that contaminants could cause an adverse effect is high) and the BPSA recommended programme for addressing extractables and leachables [5]. Assessment of leachables should be preceded by an assessment of extractables [6].

In this section, we present an approach to help end-users make a risk-based decision during selection of SUT components for a particular application and develop an appropriate process to assess operations with the highest risk that should be evaluated further for extractables. Upon completion of this risk assessment, users should obtain a data package from suppliers for the SUT that show the compounds extracted under model conditions and using standard solvents. These compounds can be assessed for toxic impact should they be present in the final dose format. In addition, if the quantity of compounds is sufficiently high, they may require further investigation under simulated or actual process conditions to identify the level at which they migrate as leachables into the active pharmaceutical ingredient (API) material. There is also a risk that leachables can impact upon the performance of processing steps, such as culture of mammalian cells, where cell growth has been shown to be impacted by compounds that have migrated from the film of the bioreactor bag [7–9].

First, the production process should be reviewed and a comprehensive list of all single-use processing materials should be made. These materials may have direct contact with the product (e.g., a product hold bag) or indirect contact with the product (e.g., storage bag for a buffer used to elute a product captured on a chromatography resin). In addition, key operating information should be documented: volumes, storage or processing times, operating or storage temperature. Other applicable information is the nature of solution being processed (i.e., pH and product stability data, if these are known to affect the extraction capabilities of plastic films). Second, a risk assessment is undertaken to evaluate if review of extractable testing is required. The key risk factors assessed are [10, 11]:

1. *Material compatibility*: Type of processing fluid and its extraction capability should be considered with organic fluids carrying the greatest risk;
2. *Proximity of component to final product*: The closer to the finished drug product, the less opportunity may be available for purification and dilution steps during the manufacturing process to reduce the concentration of leached components to ensure safe levels;
3. *Component surface area*: Dimensions can be used to determine the surface area of a component, which is important because the larger the contact area with process fluid for a given volume, the higher the exposure of the API to a leachable material;
4. *Contact time and temperature*: Duration of processing can also be used to determine exposure of the product to leachable materials, with higher operating temperatures potentially accelerating the migration process; and
5. *Pre-treatment steps*: Irradiation of single-use components increases the risk of material being extracted from plastic films, and flushing of components before use removes material generated during pre-treatment processes (irradiation, autoclaving, sanitation).

Risk scores are attributed to each identified factor, as presented in Appendix 1. Figure 6.1 presents the results of risk analysis for a general production process. Based on this graph, the operations that correspond to 80% of the total risk for extractables should be evaluated further. If components are identified that could be toxic, then further studies of leachables should be conducted.

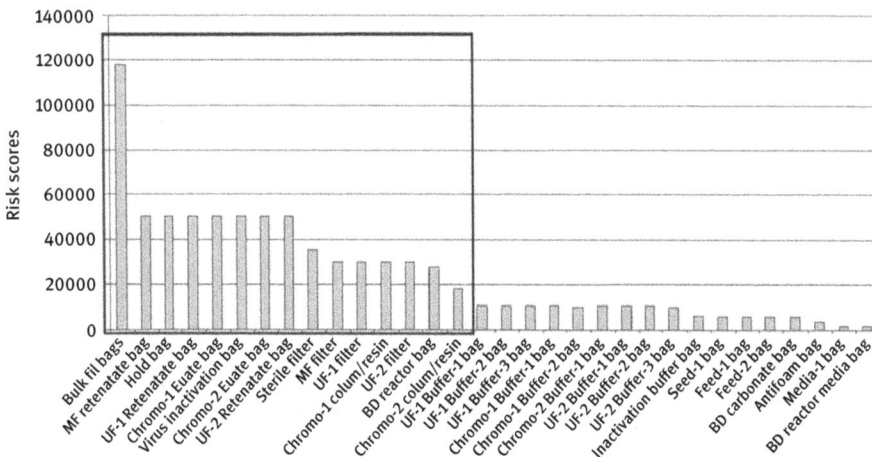

Figure 6.1: Risk scores for a general production process. Square corresponds to 80% of total risk. MF: Microfiltration and UF: ultrafiltration.

6.1.3 Profiles of extractables and leachables

Typically, SUT suppliers conduct extraction tests on their SUT products designed to characterise the extracted components and identity possible migrants. Extractable studies are usually conducted under extreme conditions (e.g., aggressive solvents, high temperatures) and with only a limited number of samples/data points gathered and analysed which represent the worst-case scenario. The degree of testing of extractables of the material component varies from supplier to supplier, though initiatives are being put forward to standardise the testing conditions of extractables [12].

End-users can make an initial assessment of the suitability of the plastic material (profile of potential leachables of SUT components) based on the analysis of the profile of extractables provided by the SUT supplier (even if only limited data are provided). An approach for this assessment is presented below. It considers the worse-case leachable profile by estimating the types and amounts of leachables generated by each SUT component during intended processing based upon worse-case conditions. This approach assists in selection of the SUT material of construction and underlines the acceptance criteria for final-assembly qualification. If data on extractables are not available for all components of the final assembly (and because significant resources are required to detect, identify and quantify extractables), the end-user may proceed directly to conducting leachables testing using simulated or actual process conditions. Depending upon the analytical capabilities and experience of the end-user, leachables testing can be conducted in-house or by using a third party.

First, based on Table 6.1, identify the type of solution group that will be processed in the final assembled system. The solution groups outlined cover the range

Table 6.1: Solution groups for extractables studies and examples of chemicals for each group.

Solution groups	Examples
Water	WFI
High pH	NaOH
Low pH	HCl
High salt (chlorine-based)	NaCl, KCl
Solvent	EtOH, isopropyl alcohol, hexane
Other	Cryopreservation media, phosphate- buffered saline

EtOH: Ethanol
WFI: Water-for-injection

of process fluids used for bioprocessing. These concern preparation of media and buffers with water-for-injection (WFI), solutions with high and low pH to mimic cell cultures (pH 4–6), and buffers used for purification processes (pH 3–11). They also include sanitisation of process components with high salt- and chlorine-based reagents, or chemicals with high or low pH, followed by flushing and rinsing with WFI. High-salt buffers with high ionic strength are also listed because they can be used during purification processes.

Not all suppliers provide comprehensive analyses of profiles of extractables covering a range of solutions, such as the ones shown on Table 6.1. Most commonly used solutions are WFI and solvents such as hexane or ethanol, which are used to create a database covering all extractable substances for a particular SUT component. The processing solution used by the end-user should be compared with the most approximate solution tested (if applicable) and this solution is assumed to behave in the same manner as the API product.

After identification of the solution group, list all materials that will be used for the final assembly of the SUT system (bags, filters, tubing, fittings, sensors, valves). Include details such as the ratio of container surface area to solution volume, storage time and temperature, because these are important factors for determination of the suitability of a container with a particular extractable profile. Also consider if the components will be (or were submitted) to any type of pre-treatment such as sterilisation or sanitation and flushing before use. Organise the information gathered as shown in Table 6.2, with one table generated for each extracted compound identified. Then, the number of material units used, the surface area tested and quantities of extracted compound can be used to estimate the amount of compound that will be generated by each SUT component and the actual scale. Then, these values can be summated to provide a total value for the final SUT assembly. It is assumed that the solution will be in contact with the entire surface area of the container for the same amount of time. Total quantity of extracted compound for the final assembly can be estimated for each compound using the formula presented in Table 6.2. Toxicity assessment can be made for extracted compounds based on the quantity of extractables calculated. For instance, a 10-l bag may be used to fill 10,000 doses, in which case the amount of extracted compounds could be averaged across all the doses for toxicity assessment. Amount of extracted compound can be compared with an analytical evaluation threshold (AET), which establishes a threshold beyond which compounds should be considered for toxic impact [11]. The AET (µg/ml or ppm) is based on dividing the safety concern threshold of 1.5 µg/day [4] by the daily dose (ml/day).

If further analysis is required, a smaller-scale assembly set-up can be used to generate leached compounds under simulated or actual process conditions to ascertain if the concentration is below the AET. If the concentration of compounds continues to pose a risk, then alternative SUT materials should be investigated.

Table 6.2: Materials used in the final assembled SUT system and associated amount of an extracted compound.

Material description	Capacity unit	Number of units	Pre- or post-treatment		Solution tested		Surface area	Minimum surface area/ volume	Conditions tested		Results (ppm)	
			Type	Conditions	Type	Volume			Temperature	Time	Value	Corrected value*
Bag container	Volume Film thickness							6:1				
Tubing, connectors	Length ID Wall thickness							6:1				
Filters	Effective filtration area							1:1				
Sensors, valves, filling needles	Total surface area ID							6:1				

* Corrected value = value × total surface area of actual unit/surface area of tested unit
Total extractabes (ppm) = ∑ (number of units × corrected value of extractables) for each component
ID: Internal diameter

6.1.4 Specification and design

As discussed above, the specification of the plastic film should enable reproducible manufacturing for upstream, downstream and aseptic bioprocessing steps, including cell culture, storage, shipping, mixing, freezing, thawing and filling applications. Key user requirements concern the physical/mechanical properties, consistent cell growth, biocompatibility, purity, robustness, gas barrier, cleanliness, and sterility of the final single-use container. All related requirements should be captured in a URS. Considerations during specification and design are listed in Table 6.3.

Single-use bag systems are off-the-shelf systems with well-understood characteristics and are offered with pre-set designs. Different systems provide different plastic materials (liners), mixing systems, sensor technologies (e.g., pH or conductivity) as well as varying bag configurations and sizes. Each supplier will provide material specification with drawings, descriptions of tubing types and connections for their design offer. If a 'bespoke' design of a bag or tubing is required, then it is preferable for the end-user to develop the design in-house and attach it to the URS. In this way, the end-user retains ownership of the design and can identify primary and secondary suppliers (if required).

Configuration of bags to meet specific requirements of the process is also possible. Even though standard bag designs are simple, tubing types and connections can be modified to suit the needs of multiple processes. Examples of different configurations of bag systems to support bioreactor operations, tangential-flow filtration (TFF) processes and fill-finish are shown in the following case study sections. The design approach presented in this section aims to standardise the types and sizes of bag assemblies as well as connection options.

Creating standard bag assembly designs helps to create interchangeable components and assemblies that can be integrated readily into any process stage, thereby simplifying logistical requirements. Standardisation of designs also simplifies procedures for preparation and filtration of solutions, which in turn reduces the burden of operator training and handling errors during operations. This approach results in a smaller portfolio of bag types, components and connections (though the number of units required may increase). The reduced number of bag-system types should simplify material inventories and, as a consequence, provide better management of stock and easier material logistics.

Figure 6.2 illustrates a schematic design of preparation, filtration and aliquoting of solutions into standard bag systems with different volumes, but with the same connectivity to support different parts of the process that have different volume requirements and which can be integrated into any processing step. Typically, the ingredients for culture media or buffers are combined with water [purified water (PW) or WFI] and mixed using an open- or closed-bag system. Once prepared, the solution is filtered through a transfer tubing set, and distributed for operations in

Table 6.3: Specification and design considerations for single-use bag systems.

Physical requirements

Geometry and capacity of the container (volume, shape, footprint, mixing, flow rate, pressure rating)
Materials of construction/film type
Permeability (product sensitivity to oxidation, pH, gas transfer, water loss)
Material compatibility (chemicals, free of animal-derived products)

Design requirements

Ports, tubing suitable for pumps, sampling, additional ports
Aseptic ports/connectors, weldable, sterilising filters

Operational requirements

Temperature, pressure, humidity, shelf-life
Sterility, closed/open system, support container type
Outer support container (transportation, dimensions, weight, footprint, functionality)
Transportation (size, handling, shipping conditions)
Storage (plant storage and movement, footprint suitability, safety aspects)

Process control/monitoring

Compatibility with existing support equipment such as pumps, sensors, support container

Quality and regulatory requirements

Biocompatibility*
Mechanical properties (tensile strength, elongation break, seal strength, air-leak test)
Gas/vapour transmission (ASTM D3985: Oxygen; ASTM F1249: Water vapour)
Compendial physicochemical properties
TSE/BSE status*
Analysis of total organic carbon*
pH and conductivity analysis of rinsate*
Extractables and leachables*

Chemical compatibility
Protein adsorption studies
Endotoxin testing*
Sterilisation validation*
Integrity of container closure
Particulates*
Bacterial ingress (bags) (USP 71 Sterility tests; EP.2.6.1 Sterility) and challenge (filters) tests (HIMA/ASTM F 838-05 Bacterial challenge)

---

* Details of tests and guidelines outlined in Table 5.1
ASTM: American Society for Testing and Materials
BSE: Bovine spongiform encephalopathy
HIMA: Health Industry Manufacturers Association
TSE: Transmissible spongiform encephalopathy
USP: United States Pharmacopeia

a)

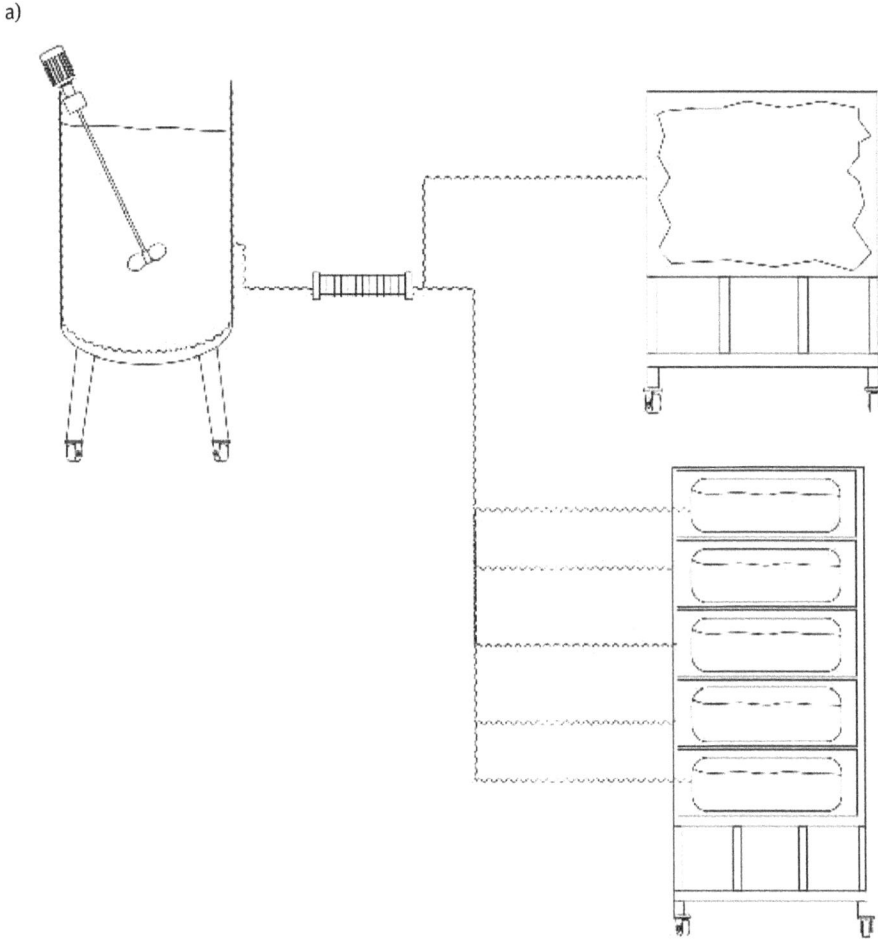

Figure 6.2: Detailed single-use bag system: (a) solution preparation followed by sterile filtration split into various connections and into various solution storage systems; (b) standard bag designs suitable for varying solutions volume requirements, parts and connectors; and (c) buffer bags connected into a control system for chromatography, and outlet connections to waste bags and intermediate-product hold bags. (continued on p. 76)

either a large container, or aliquoted into smaller bags. In the example shown, one tubing set is used to connect to all of the hold bags, which helps to reduce costs. If separate tubing sets were used to connect the preparation bag to each hold bag, then the number of filters required will increase. These are usually expensive components, so reducing the number used helps to control costs, particularly in facilities in which many solutions will be prepared and aliquoted into hold bags. The transfer tubing set does not need to be connected to all bags before aliquoting commences. However, the design should incorporate the ability of the tubing set to allow for aseptic

b)

Luer
MPC
Sterile
Tri-clamp

c)

| WFI | Regeneration | Elution | Wash 1 | Wash 2 | Equilibration |

Product

Intermediate product   Sanitation   Waste

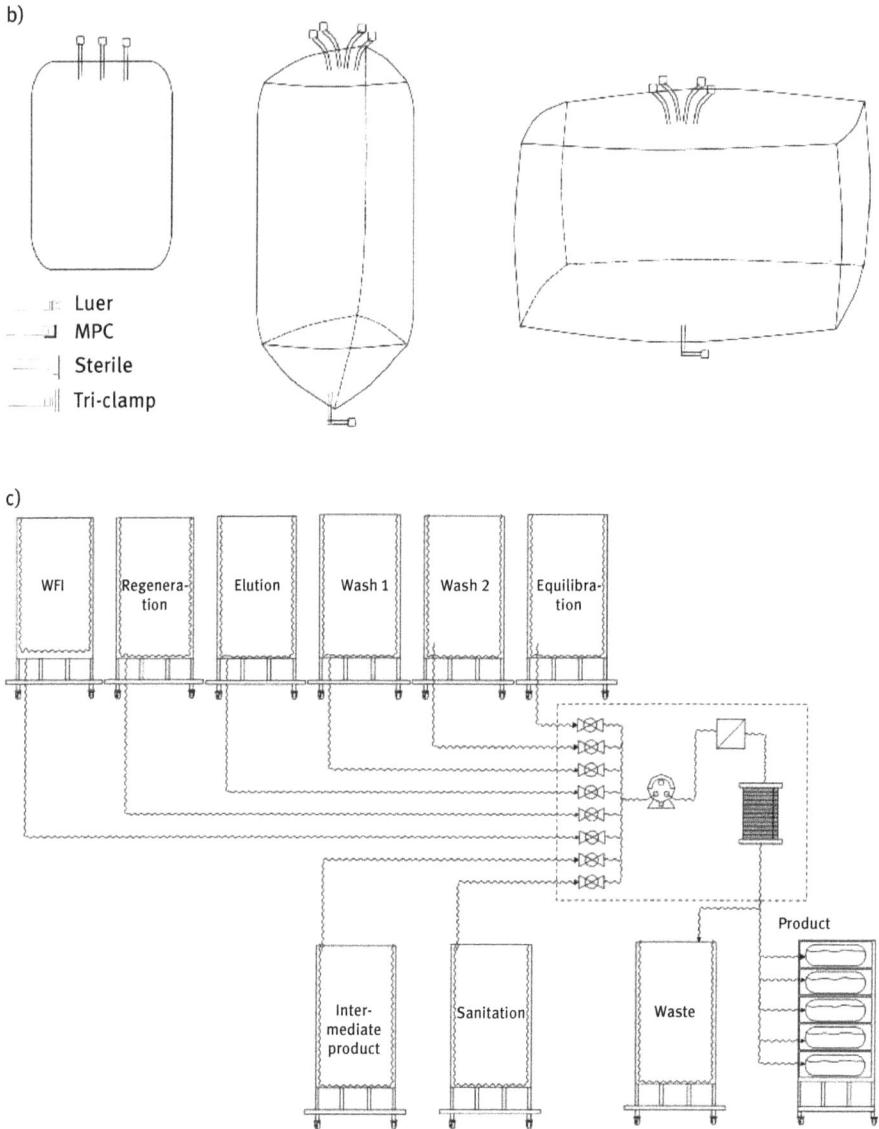

Figure 6.2 (continued)

connection and disconnection between bags using sterile aseptic connector components or tubing welding. Once filled, the hold bags must be transferred into the final
point of use and integrated into the unit operation. Chromatography and TFF operations require greater volumes of (sometimes) different solutions, as well as collection
of the waste generated. These standard configurations can be targeted for these

operations or can be used for intermediate production hold or storage of the final bulk drug substance.

An example of the number of different types of bags required in a facility using general process described in Section 6.1.2 is shown in Table 6.4 for two cases: (i) if the volumes of solution required are matched to the closest corresponding bag size; and (ii) if the size of bags are rationalised down to a standard range. In the first instance (before standardisation), 9 bag sizes and types are required at different stages of the process to give a total of 68 bags. This estimate does not include the bags required to remove waste from the process. However, it can be assumed that, if these were included, then this value will increase significantly. Reviewing the sizes that are required, it is possible to utilise the 'turndown ratio' (the operational range of the bag, defined as the ratio of the maximum capacity to minimum capacity) to reduce the number of bag types required. For example, 250-l bags can handle working volumes of 100–250 l, so they could be used instead of 100-l or 200-l bags. Alternatively, multiple bags could be used to replace larger bags, for example, using 2 × 1,000-l bags as oppo sed to 1 × 2,000-l bags. If we apply this rationalisation, it can be seen from Table 6.4 that the number of bag types can be reduced to 4 (after standardisation), with only a small increase in the total number of bags required to 70. Another advantage of the decrease in the range of bag types used is that the number required of each bag type increases. This approach provides a stronger position when negotiating with bag suppliers, which can be used to reduce the cost per bag. This is a useful advantage given that, because 1,000-l bags are being used instead of 500-l bags, the cost per bag might increase. However, it is not typical for bag cost to double between scales, so negotiations on price might allow for the total cost of all bags to not increase significantly.

Technical requirements for bags should also be accounted for, such as: the support system required (e.g., mixing, pH measurement, conductivity); and if a minimum number of ports or sterile connections will be required; storage requirements; if the bag will be submitted to stress (e.g., low/high temperatures, high and low pH solutions, mixing and pumping). However, as previously mentioned, tubing sets can often be used to configure standard bag designs.

6.1.5 Qualification of final bag assembly

The Food and Drug Administration (FDA) states that qualification of container closure systems should be based upon '*establishing compatibility and safety of the liner. . .*' [13]. The container material should not adversely affect the quality of the product, or allow diffusion across the container into or out of the preparation. Other considerations may also include '*characterisation for solvent and gas permeation, light transmittance, closure integrity, ruggedness in shipment, protection against microbial contamination. . .*' [13]. All bag systems do not have to be qualified

Table 6.4: Number of bags required at different stages of a general 4 × 1,000-l production process before and after standardisation.

Before standardisation Bag volume (l)	Number of items										Total
	Upstream	Downstream									
		Buffer preparation	Hold	UF-1	Chromo-1	Virus inactivated	Chromo-2	UF-2	Fill		
2	1										1
20	4					1			6		11
50	1										1
100	8							3			11
200	8	1	2	2		1	1	1			16
250				4	1						5
500		1		1	5		6	1			14
1,000	4	1	1	1	1						8
2,000		1									1
Total	26	4	3	8	7	2	7	5	6		68

After standardisation										Total
2	1									1
20	6							1	6	13
50										0
100										0
200										0
250	16	1	2	6	1	1	1	4		32
500										0
1,000	4	4	1	2	6	1	6			24
2,000										0
Total	27	5	3	8	7	2	7	5	6	70

to the standard laid out for container closure systems. However, the principles outlined above are a useful guide on which to review each application of bag systems
within the process.

The most important factors to qualify the final bag assembly for a manufacturing environment are integrity, maintenance of sterility and cleanliness, as well as
the stability and purity of the API. These factors should be qualified by the end-user
using complete equipment and three different lots of SUT that comprise the final
bag- assembly system run under normal operating conditions (i.e., solutions, temperature and hold times). Characteristics of the plastic film (e.g., exchange of gas
and moisture) and the physical strength of the bag components should be certified.
If one bag has been validated for use with an API product (under certain processing
conditions) and is discontinued, then the new replacement bag system must be re-
validated. Dual suppliers of different bag assemblies/components should be validated in parallel with the primary option. This strategy minimises costs and reduces
the risks associated with the security of the supply chain. All technical modifications to the final assembly design or parameters should be controlled and tracked
under a change-control process.

Qualification of system integrity ensures that the system set-up is maintained
through processing and that it has the appropriate physical and mechanical properties to avoid loss or contamination of the product. If a large-scale system holding
large volumes of valuable product is to be validated, and the risk to the final drug
product is low, then a scale-down model of the final assembly can be utilised. In
this case, the equivalent surface-area calculations (such as the ones shown in
Table 6.2) could be used to scale-up the measurement of particulates and leachables resulting from the system.

Table 6.5 shows the risks associated with container systems and parameters
that require testing if systems are used for three major applications: cultivation of
cells; preparation and storage of media or buffers; intermediate hold or formulation
and storage of the final product. Different risks are associated with various applications, which need to be qualified independently. Based on these risks, a general
overview of the type of trials and tests to be conducted (as well as parameters that
must be qualified) is outlined. Considerations for each particular system used for
each specific application should be based upon a risk assessment tailored for a particular product.

For all applications, the physical and microbiological integrity of the final assembly must be guaranteed, so the robustness of container seals, connections, and sampling ports must be tested to ensure that contamination does not occur. Usually, a
100% leak test is sufficient to guarantee no breaches or leaks pre- and post-process-
ing, or during storage or shipping operations. Bacterial ingress tests may be required
when the bag system is used for storage of solutions. Tests need to prove that the
bag system maintains sterility throughout storage time and under storage conditions.
Robustness is also important if larger volumes are processed, lower or higher

Table 6.5: Risks, testing and qualification parameters for bag systems according to their application: cultivation of cells; preparation and storage of media and buffers; formulation and storage of intermediates or bulk drug substance.

Application	Risks	Testing	Parameters
Bags used for preparation and storage of media/buffer	Bioburden and particulate control Leachables (depending upon where the bag is used within the process – see Section 6.1.2) Properties of the solution change during storage (e.g., homogeneity, conductivity, pH)	Certification from SUT supplier Mixing studies Pre-use sterilisation method (gamma irradiation or autoclaving) Hold time studies	Temperature Capacity (volume) Duration Robustness: bags and connections (100% leak test) Transport Manipulations Leachables and extractables Mixing
Cell cultivation bags	Maintenance of sterility (robustness of bag, connections) Release of particles (mixing, pumping, filtration) or leachables (from film) that may interfere with growth/productivity Limited cell growth due to limited mixing or mass transfer	Sterility testing (media hold - see Section 5.2.2) Growth performance Robustness tests (container and connections) Analyses of extractables and leachables Pre-use sterilisation method (gamma irradiation or autoclaving) Mixing studies and characterisation of mass-transfer properties	Duration of culture Temperature Agitation and mass-transfer properties Pressure Volume Robustness of bag and connections (100% leak test) Leachables and extractables

Table 6.5 (continued)

Application	Risks	Testing	Parameters
Intermediates and bulk drug substance hold/storage bags	Maintenance of sterility (robustness of bag, connections) Release of particles (mixing, pumping, filtration) or leachables (from film) that may affect product purity Properties of product change over time (e.g., concentration)	Sterility testing (media hold - see Section 5.2.2) Robustness tests (container and connections) Mixing studies Analyses of extractables and leachables Pre-use sterilisation method (gamma irradiation or autoclaving) Hold time studies	Duration Temperature Agitation Pressure Volume Robustness of bag and connections (100% leak test) Leachables and extractables Mixing API properties

temperatures are utilised, sampling manipulations are required and transportation of materials is foreseen. Some bioprocessing bags require validation for storage at lower temperatures, such as -20 °C or lower. Integrity of plastic materials submitted to sterilisation procedures (e.g., irradiation, autoclaving) must also be assessed. Release of particulates generated during agitation of solutions, transfer using pumps and filtration, as well as substances that may leach from plastic film that may interfere with cell growth or API purity must be quantified. This strategy is based upon a risk assessment and must be validated during qualification of the process.

The impact on API can be measured by loss of potency due to leached materials from the plastic liner component. Alternatively, properties of the liner could induce degradation, precipitation, changes in pH or discoloration of the API. As mentioned above, these features should be detected during qualification studies. If different bag sizes are to be used, a leachables test should consider the surface-to-volume ratio of the test container–solution combination, as shown in the extractables assessment presented in Section 6.1.2 and 6.1.3.

For preparation of media and buffers, the end-user relies solely on certification and testing provided by the SUT supplier related to levels of particulates and bioburden, as well as sterility assurance. The end-user will require a certain level of cleanliness of the SUT materials used for preparation of media and buffers to control the levels of contamination that may enter processing cleanrooms.

If sterility must be qualified, this can be achieved by undertaking a media hold (see Section 5.2.1) as described in FDA guidelines for qualification of aseptic processes [14]. This qualification should be undertaken using equipment employed for normal operations and under standard processing conditions. In the case of bags used for cell growth, normal duration of culture should be tested with maximum fill volume and maximum stirrer speed under normal operating conditions such as temperature and pressure (see Section 6.2 for more details regarding qualification of a SUB system).

6.2 Case study 2: Single-use bioreactor

A SUB system comprises a support structure in which a cultivation bag is contained and connected to an integrated monitoring/control system. This support structure provides the required agitation level *via* a mechanically or magnetically driven stirring mechanism, or *via* an oscillating rocking movement. Temperature is a key process parameter and is maintained by a jacketed vessel support, or by a heating element contained within the support platform. The cultivation bag is a biologically inert pre-sterilised plastic polymer. It contains various addition ports (for feed addition, antifoam, acid and base, inoculum), sampling ports, and vents with integrated filters. In addition, culture bags come equipped with disposable online pH and dissolved oxygen (DO) probes, or ports designed specifically for insertion of re-usable probes. These bags have different configurations depending on the intended use (e.g., microbial/mammalian cell culture, microcarriers, perfusion mode). Gases can be supplied *via* headspace addition or a sparger design configurations incorporated into the bottom of the bag.

SUB benefit from elimination of CIP and reduction of SIP activities, which reduces set-up time and results in faster times for batch turnaround in addition to a reduced risk of cross-contamination across batches. SUB are limited to a 4,300-l scale (working volume, 3,500 l) for well established applications of mammalian cell cultures. More recently, systems designed specifically for microbial applications have been placed on the market with scales ≤300 l. Besides scale limitations, the main technical limitations of this technology are:

- Pressure limits of the bags (0.5–0.7 psi), which in turn limits rates of gas flow and oxygen transfer;
- Capacity for agitation and sparging technology to deliver sufficient levels of mixing and oxygen transfer to the culture; and
- Poor thermal conductivity of the plastic bag, which may interfere with the ability of the system to maintain temperature and/or remove heat generated during culture.

The complexity of the operations that take place within the bioreactor means that care should be taken when selecting a system because, once the decision is made, substitution without a major impact upon operations will be difficult. SUB performance is a key driver of the productivity of the overall process. Performance differences will probably be seen across SUB designs. Careful understanding of existing requirements and taking into consideration potential scale-up is advisable to reduce the risk of reaching limits that restrict the ability to meet future manufacturing targets.

6.2.1 Selection of single-use bioreactor technology

Selection of SUB technology should involve technical feasibility and economical evaluations of all available and applicable technologies. Significant work has been published concerning the parameters to be evaluated during the feasibility assessment and scale-up of SUB technologies [15–18]. The overall consensus is that a trial should be conducted using a well understood process with a model cell culture/ product. Special attention should be given to the control of critical processing parameters, such as oxygen transfer, mixing time and power input [15, 16, 18]. If possible, different scales should be compared to assess the scalability of technology.

This section presents an example of a combined technical and economic risk-based approach for selection of different SUB technologies. Table 6.6 presents a summary of different SUB technologies alongside key considerations for selection

Table 6.6: Summary of requirements and risks associated with different SUB technologies used for small-scale research and development activities, and pilot-scale clinical manufacture.

Application	Small-scale research and development			Pilot-scale clinical manufacture		
	Bench-top stirred tank glass bioreactor	Bench-top stirred tank plastic bioreactor	Culture bag in rocking system	Stainless-steel stirred tank bioreactor vessel	Stirred tank bioreactor support system for culture bag	Culture bag in rocking system
System details	Bench-top stirred tank glass bioreactor	Bench-top stirred tank plastic bioreactor	Culture bag in rocking system	Stainless-steel stirred tank bioreactor vessel	Stirred tank bioreactor support system for culture bag	Culture bag in rocking system
Description	Glass vessel	Hard plastic vessel — Inserted into a self-contained support control system / Designed to use existing control system	Stainless-steel platform with single-use flexible plastic culture bag	Stainless-steel vessel	Stainless-steel support with single-use flexible plastic culture bag	
Results from trial	Final volume, growth levels, product yields and quality[1]					
Materials and consumables	Cleaning reagents, water and steam (autoclave), replacement parts kit	Consumable parts: culture vessel or bag, probes (if not re-usable)	Consumable parts: culture vessel or bag, probes	Cleaning reagents, water and steam, replacement parts	Consumable parts: culture bag probes (if not re-usable)	Consumable parts: culture bag probes (if not re-usable)
Capital cost	Cost of control system and vessels	Cost of control system	Existing control system may be used (if applicable)	Cost of support and control system	Cost of vessel and control system	Cost of support system and control system

Table 6.6 (continued)

Application	Small-scale research and development				Pilot-scale clinical manufacture			
FTEs (days)	Details to be completed for a particular process and tested during trial							
Complete process duration (days)								
Support systems required	Autoclave (steam), PW, cooling water and gases. If re-usable probes to be used, these require autoclaving (steam)	Cooling water or external cooling system and gases	Gases	Control tower, cooling water, gases. If re-usable probes to be used, these require autoclaving (steam)	Steam for autoclaving and SIP, PW, cooling water and gases	Cooling water or external cooling system and gases	Gases	If re-usable probes to be used, these require autoclaving (steam)
Main strength	Established technology for small-scale bench-top work	High throughput analysis (small-scale and fully disposable)	Design and operation very similar to established bench-top stirred tank glass bioreactor. Does not require cleaning and sterilisation	Established/mature technology. Maximum working volumes available ≤500 l	Established technology for pilot-scale manufacture	Does not require cleaning and sterilisation between batches. Similar principle to established stirred tank bioreactor technology. Maximum working volumes available ≤300 l (microbial) and 4,300 l (mammalian)	Established/mature technology. Maximum working volumes available ≤500 l	

Main risk[2]	Reliance on utilities Requires cleaning and autoclaving between batches	New/immature technology Not fully tested (performance and scalability)	No universal scale-up methodology established Embedded probes may drift in terms of accuracy during operation	Reliance on utilities Requires cleaning and sterilisation validation between batches	No universal scale-up methodology established Embedded probes may drift in terms of accuracy during operation	Unproven performance using traditional or alternative mixing and sparging technologies Sterilisation of re-usable probes Embedded probes may drift in terms of accuracy during operation
						No universal scale-up methodology established Embedded probes may drift in terms of accuracy during operation
Risk mitigation	Use hybrid system with all ancillary equipment being SUT/irradiated	Undertake a trial with technology by the end-user Test scalability of system	Undertake a trial with technology by the end-user	Use hybrid system with all ancillary equipment being SUT/irradiated	Undertake a trial with technology by the end-user	Undertake a trial with technology by the end-user

[1] To compare with specific parameters and results tested and obtained during a trial of a specific technology (for a specific process).

[2] Specific risks associated with a particular process should be added by the end-user and tested during the trial.

FTE: Full time equivalent.

of these systems for different applications, whether small-scale research and development, or pilot-scale production of clinical material. This type of analysis should be applied to the different applications required by the end-user because the costs of each technology (as well as time savings in terms of processing) may vary. The support systems required by each technology give an indication of fit to an existing facility and highlight changes that may need to be undertaken to adopt the technology. For all the technologies presented, the importance of the analysis of risks and generation of an appropriate risk-mitigation strategy are highlighted. This strategy also gives visibility on additional systems that have to be in place for a specific technology to be used in the future, and highlights critical points that must be tested during a trial. This assessment should be undertaken by Subject matter experts.

Leachables from the plastic material of the culture bag should be considered as early on in the process as possible because they may affect cell growth and/or productivity [7–9]. Leachables could also have a significant impact on biological products that are cell-based systems (e.g., stem cells or live attenuated viruses) because there are few purification steps that could be used to remove toxic substances to safe levels for injection into patients. More details regarding preliminary analysis and qualification of leachables during process qualification are shown in Section 6.1 with regard to all types of bag systems (including for cell cultures).

6.2.2 Specification and design of single-use bioreactors

At the start of SUB development, user requirements covering cell growth, productivity, robustness, reliability and security of supply are defined. Table 6.7 presents an example of a generalised URS with key considerations for a SUB system. The URS should start with an overview of the SUB purpose (type of cells grown, product type, clinical/ development, process area) and more specific details of type of equipment used, components and capacity. A process description, such as that given in Figure 6.3, provides a useful summary of the key steps within SUB operation, and can be coupled with the URS to describe the purpose of the system to be reviewed by the supplier. Figure 6.3 presents all activities undertaken in the SUB system as a sequence, and they are broken down into individual process steps according to their different aims. The equipment and materials required for each step, solutions entering and leaving the system (e.g., product streams, buffers, compressed air and samples) and key process parameters to be considered are listed. The operational requirements established in the URS describe control parameters of the process and control–software interfaces that support such control. Description of the SUB culture should specify the materials of construction, general requirements of bag design (addition and sampling ports), as well as relevant quality and certification requirements. The URS should also outline additional documentation,

Table 6.7: URS for a SUB system.

1) **Overview**	
Key features of equipment and components	The purpose of this document is to describe user and functional process requirements for a SUB skid, utility skid and ancillary equipment to be used within a GMP facility
Key quality deliverables to ensure sufficient information to enable equipment to be designed and qualified to meet requirements	This document is intended to specify: – Operating performance of a system – Construction requirements – Quality and regulatory requirements – Provide baseline for validation activities – Identify the scope of vendor supply

2) **Objectives**	
What will the equipment be used for/applications and its capacity/ scale of operation	The SUB will be used for cell growth (type of cells to be grown) in a controlled environment to a quantity/concentration/viability suitable for harvest/production of (product type) to a determined level. The SUB will be located in the process area 'X' and will be operated in batch/fed-batch/perfusion mode. The final culture will be harvested via centrifugation/filtration and frozen. Culture can also be discarded to biowaste The system will be used for cell cultures with volumes up to 'X' l

3) **System and process definition**	
Schematic for: – Process steps (include process descriptions and flowcharts as appropriate) – Main pieces of equipment	See Figure 6.3 for process description List of main components: – Bioreactor skid (jacketed stainless-steel vessel containing a control unit with integrated agitation, gases delivery, pressure, and additional pumps for antifoam addition and pH control – Temperature control unit – Control system – Ancillary equipment (condenser)

4) **Operational requirements**	
4.1) Capacity	The system must allow for cultivation of (microbial/mammalian) cell cultures with volumes between 'X' and 'X' l
4.2) Process requirements Key requirements that will form the basis of IQ/OQ procedures	The system must be able to maintain: – Key process parameters within set-points, gather and record these data, and make batch data available post-processing for printing and data trending – Temperature control – the process temperature should be measurable and controllable to a range of e.g., 20–42 ± 0.5 °C for a period of up to 'X' h. The unit will be able to cool down to e.g., 10 °C

Table 6.7 (continued)

	– Control of agitation speed
	– DO control
	– pH control – pH probes and acid/base pumps for addition and control of pH
	– Gas flow rates
	The system, in conjunction with the bioreactor bag, must maintain mono/asepsis for ≥'X' days
4.3) Control and monitoring of the process List critical process parameters, set-points or measurement range, location of measurement (vessel, jacket) and accuracy/ tolerance of measurement and duration	Control parameters, measurement range, accuracy/tolerance and duration: – Temperature vessel – e.g., 20–42 ± 0.5 °C – Temperature jacket – Temperature utility – Agitator speed – pH – DO – Air flow – O2 flow – CO2 flow – Media mass flow – Pressure – Weight measurement (vessel volume) The instrumentation must be able to undergo calibration or be supplied calibrated
5) Control requirements	
5.1) System control	The system must be able to: – Store sufficient data – Allow configuration of all equipment parameters/recipes The control system will be the means of: – Monitoring and controlling the equipment (bioreactor) – Viewing and resetting alarms – Printing batch information
5.2) Quality requirements	Instrumentation and control systems should comply with: – 21 CFR Part 11 – Electronic records, electronic signatures – GAMP® 5 – Guide for validation of automated systems – EU Annex 11 Volume 4
5.3) Security requirements	Individual users with different user ID and password Individual users with different access levels (Table 6.8)

Table 6.7 (continued)

5.4) Data and reporting	All parameters will be stored, trended and available for viewing and printing
	Control system shall be able to generate a batch report
5.5) Power failure	In the event of a power failure, the system configuration shall be maintained and should also recover data and restart automatically
5.6) Stop	The equipment shall be provided with an emergency stop button (in compliance with CE)
5.7) Faults and alarms	Detailed list of system alarms (Table 6.9)
6) Materials of construction	
6.1) Support frame	The support frame and associated equipment surface materials must be suitable for cleaning and have sufficient access for cleaning
	Materials of construction should be suitable for an ISO 8 cleanroom environment
6.2) Single-use bag Specific design requirements should be specified	The bag system should include, as a minimum, the following: – Temperature probe port/system – One/two pH probe port system – One/two DO probe port system – One antifoam sensor port/system
	One pressure transmitter (bag pressure monitor) One sample port suitable for taking multiple samples
	Additional ports for: acid, base, antifoam, culture media, inoculum, process gases, harvest lines, feed line, inducer solution
	Agitator system
	Single-use bag and associated tubing (product contact parts) should have as a minimum: – USP Class VI Biological test for plastics – ISO 10993 or USP 87/88 - Biological evaluation of medical devices – ISO 11137 Sterilisation of healthcare products - Assurance of sterility by gamma irradiation (if required) – EP 5.2.8 - TSE/BSE-free manufacture
7) Quality and regulatory requirements	
Describe supplier testing requirements, delivery, qualification	The equipment shall be designed to comply with the requirements of EMEA and FDA GMP

Table 6.7 (continued)

	CE certification
	FAT
	Instructions on installation, operation and maintenance
	Instrumentation and control systems should comply with standards specified in Step 5.2
	Support frame and associated equipment such as a single-use bag should comply with industry standards specified in Step 6 (Materials of construction)
8) Interface requirements	
Describe equipment interface requirements	Type of electrical supply
	Bioreactor bag and tubing systems must interface with a range of welders and sealers (if applicable)
	Autoclavable parts (if applicable)
9) Functions	
Describe main equipment functions	The system should operate with minimal involvement of the operator
	The system shall be operated locally by means of an operator interface terminal
	The system control loop can be configured to carry out P&ID control, set-point control, cascade control and timer control
10) Utilities	
Describe equipment utility requirements	The operation of the system must be able to function with the available utilities:
	– Electricity
	– Process gases (O_2, N_2, CO_2) at 'X' barg
	– Process 'X'% gas/air mix at 'X' barg
	– Clean steam and non-clean steam at 'X' barg
	– PW at nominally 'X' barg
	– Chilled water supply at 'X' barg and 'X' degC
	– Chilled water return at 'X' barg
	– WFI at nominally 'X' barg
	– Towns water at 'X' barg
	– Instrument compressed air at 'X' barg
	– Clean compressed air at 'X' barg
	– Biowaste drain
11) Maintenance	
Describe expected hardware and software maintenance support required	Describe expected hardware and software maintenance support required

Table 6.7 (continued)

---

12) Environment

Provide details of the physical environment in which the bioreactor will be operated	Layout (attach drawings with dimensions of room and equipment) Room classification Cleaning requirements (specify agents used)

13) Additional information

Additional information to be sent to end-user for review and approval	Before delivery the supplier shall provide the following information: – List of requirements for SAT/IQ/OQ (if supplier is undertaking it) such as bill of materials, utilities required, spare parts list and protocols for review and pre-approval – Maintenance instructions and schedule of equipment and software – P&ID – Control and equipment assembly drawings and dimensions – Control software disc file – FAT – tests on: control, hardware and operation – The supplier shall provide the following documentation/copies of: – Operation, installation and de-commissioning instruction manuals – Quality and project plan (available for review) particularly for bioreactor bag supply – Materials of construction certification – Calibration certificates

14) Supplier support

Describe support activities required after delivery of system	Post-delivery the supplier shall provide the following support: – Start-up support – Training – Technical support – Preventative maintenance – System improvements/updates

CE: European conformity
EP: European Pharmacopeia
FAT: Factory acceptance test
GMP: Good manufacturing practice
IQ: Installation qualification
ISO: International Organization for Standardization
OQ: Operational qualification
P&ID: Piping and instrumentation diagram
USP: United States Pharmacopeia
SAT: Site acceptance test

Materials	Equipment	Process aim and description	In/Out	Key process parameters
Sterile culture bag Tubing, connectors, probes and ancillary equipment	SUB system Autoclave Laminar flow	**Material preparation and set-up** Preparation of materials/tubing sets/probes to be sterilised Sterilisation of non-sterile materials/tubing/probes sets by irradiation or autoclaving Set-up and assembly of sterile culture bag to SUB stainless-steel support and connection to pre-sterilised tubing sets/ancillary equipment	In: Gases (air, N_2)	Sterilisation process conditions Visual inspection of culture bag and all tubing sets
Chemicals Media Filter	SUB system Balance Stirrer platform Pump Welder Sealer Integrity tester	**Media preparation, transfer and equilibration** Preparation of solutions and filter sterilisation Transfer of media into culture bag, equilibration under operating conditions and re-calibration	In: Sterile media Gases (air, O_2, N_2, CO_2) Acid/base	Filter integrity test Temperature, pH, aeration, mixing, pressure and so on Duration of equilibration Re-calibration of pH and DO probes
Inoculum strain Feed media suspension Induction solution	SUB system Laminar flow Welder Balance Pump	**Production in SUB** Inoculation of SUB Maintenance of culture conditions and bag integrity Addition of feed media suspension and induction solution for product expression	In: Inoculum Sterile media Gases (air, O_2, N_2, CO_2) Acid/base Induction chemical	Temperature, pH, aeration, mixing, pressure Feed additions/rate Induction strategy (solution, concentration, temperature and so on) quantity and duration Duration of culture
Sample collection bags/vessels/ containers	SUB system Separation method Cell density and viability measurements Metabolites analyser Fridge/freezer	**Monitoring and sampling culture** Monitoring process parameters Collection of samples for analysis	In: Sterile media Gases (air, O_2, N_2, CO_2) Acid/base Out: Culture samples	Temperature, pH, aeration, mixing, pressure and so on Sample collection frequency, quantities, preparation and storage conditions Measurement of cell density, viability of cells and metabolites
Harvest collection bags/vessels/ containers	SUB system Pump Primary recovery equipment Cold room/freezer	**Harvest/transfer culture** Harvest of culture Processing and storage of culture	In: Gases (air, O_2, N_2, CO_2) Acid/base Out: Culture	Harvest conditions Storage conditions
Decontamination and cleaning agents Waste collection bags	SUB system Pump	**Decontamination and cleaning** Decontamination and cleaning of culture bag and ancillary equipment used for processing	In: Cleaning agents Water Out: Used bag waste Liquid waste	Time and concentration decontamination and cleaning agent

Figure 6.3: Process description for SUB unit operation.

support and services required from the SUB supplier such as design, training, validation, and maintenance. Table 6.8 shows an example of a potential set-up of security levels and Table 6.9 presents an example of critical alarms. These tables support the information shown in the URS of the SUB system (Table 6.7).

Table 6.8: Security levels.

Access levels	Operator	Supervisor	Maintenance	Engineer	Administrator
Ability to operate and monitor the system	×	×	×	×	
Start/stop system/ data acquisition	×	×	×	×	×
View and acknowledge alarms	×	×	×	×	×
Calibrate DO, pH probes and pumps (ancillary equipment)	×	×			×
Reset container/ bag weight	×	×			×
Enable/disable alarms		×			×
Change analogue values (sequences and recipes)		×			
Place system mode of operation		×	×	×	×
View analogue values		×	×	×	×
Access calibration and maintenance functions			×	×	×
Change P&ID control parameters				×	×
Change P&ID set- points and modes		×			×
Remote device control	×	×	×	×	×

Operator: Run system on a day-by-day basis
Supervisor: Change system within validated limits
Maintenance: Carry out periodic maintenance/calibration of system
Engineer: Change GMP parameters as necessary
Administrator: All activities except system operation

SUB systems are off-the-shelf systems with well-understood characteristics and pre-set designs. Different systems provide different product contact plastic materials,

Table 6.9: Critical alarms.

Alarm/information message	Critical alarm	Light illumination	Information message	Response	
				Interlock	Procedural
Emergency stop	×	×	×	×	×
Control platform/unit communication	×		×	×	×
Control platform/unit power fault	×		×	×	×
High/low operational parameter1 value set- point	×		×	×	×
Deviation of operational parameter1 value set-point	×		×	×	
Jacket water supply fault	×		×	×	
Gases supply fault	×		×	×	
Rupture disc failure	×		×	×	×
Motor fault	×	×	×	×	×

[1] Temperature (bag, jacket, temperature control unit), pH, DO, agitation speed, gases supply flow, bag container pressure, bag container weight

mixing systems, sensor technologies (to measure, for example, pH and DO) and varying bag configurations and sizes for cell cultures. Each supplier will provide material specifications with detailed assembly drawings, parts lists and references for materials.

Figure 6.4 presents an example of a SUB process flow diagram (PFD) that provides further detail for the design requirements of the operation for a specific purpose. A potential arrangement for a pilot-scale SUB operation and all ancillary equipment for a fed-batch culture operation can be observed. Several utilities and solutions will pass through the system boundaries such as air, harvest culture, and addition of acids/bases acids/bases. The system should be designed to have sufficient inlets/ outlets to handle process requirements. The PFD provides a sufficient level of detail to portray the intent of the system, and can be coupled to a URS as part of a package of specification documents to be passed to potential suppliers for review.

Single-use bags have many features (probe ports, sparger, liquid transfer lines) that can be used to suit the needs of multiple processes. Many suppliers provide pre-established designs for culture bags with pre-assembled tubing and filter assemblies required for a specific use (microbial or mammalian cells) that add

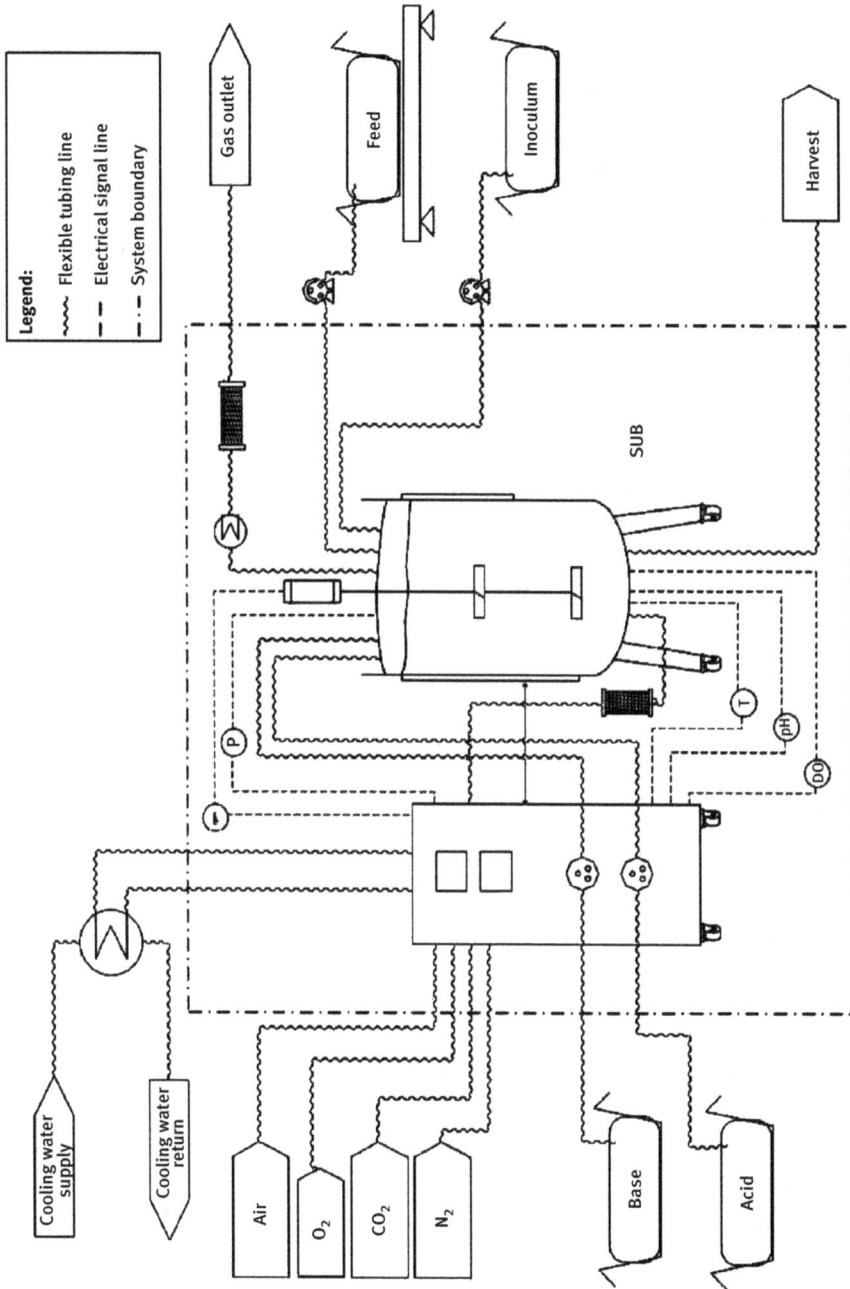

Legend:

〰〰 Flexible tubing line
‒ ‒ Electrical signal line
‒·‒ System boundary

Gas outlet

Feed

Inoculum

Harvest

SUB

Cooling water supply

Cooling water return

Air

O$_2$

CO$_2$

N$_2$

Base

Acid

Figure 6.4: SUB PFD.

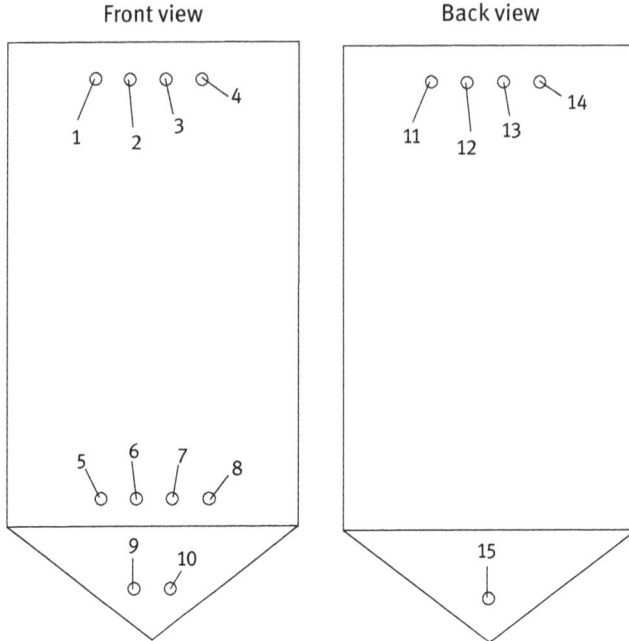

Port	Description	Tubing set (type, length, ID, OD)
1	Media addition	1 m CFlex 374 3/8 "ID × 5/8" OD
2	Pressure sensor	N/A
3	Base addition	3.9 m CFlex 374 ¼ "ID × 7/16" OD
4	Acid addition	3.9 m CFlex 374 ¼ "ID × 7/16" OD
5	Temperature probe	N/A
6	Probe ports	
7		
8		
9	Bottom drain harvest port	3 m CFlex 374 3/8 "ID × 9/16" OD
10	Sampling ports	N/A
11	Antifoam	3.9 m CFlex 374 1/8 "ID × ¼" OD
12	Feed addition	3.2 m CFlex 374 ¼ "ID × 7/16" OD
13	Exhaust gas	N/A
14	Inoculum addition	3.2 m CFlex 374 ¼ "ID × 7/16" OD
15	Inlet gas sparger	N/A

Figure 6.5: Example of standard culture bag.

flexibility and connectivity to the design of the SUB bag. The bag design is likely to be owned by the supplier and alterations to the standard design or customisations must be agreed with the supplier. Alternative design configurations can be obtained only by changing the inlet/outlets to the system.

Figure 6.5 is a schematic design of a standard culture bag in which different types of tubing are associated with each port as well as connection types. In this figure, ports are labelled and suggestions are presented for connections to ancillary equipment such as feed, base, acid, antifoam, inoculum, and probes. An example of schematic drawings of the design of individual components that comprise a sub-system to be connected to a SUB bag is shown in Figure 6.6. This sub-system is to be used for addition of feed media through port 12 of the culture bag (Figure 6.5). It can be broken down into three individual components that must be connected together to form the final sub-system assembly. This sub-system allows for sterile filtration of the feed solution into the pre-sterile addition bag, disconnection of the filter for integrity testing, and specification of suitable tubing welding to port 12. The first component is a standard pre-sterilised addition bag of an appropriate size to hold feed media solution. Components 1 and 2 are made of silicone tubing suitable for use within a pump, cut to different sizes and with appropriate connections such as MPC for connection with a standard bag. Component 3 is formed of weldable tubing of equivalent size [internal diameter (ID) and outer diameter (OD)] to the one provided by port 12, and a MPC for connection with a standard bag. Vent filters have been placed in all components in open-ended tubing (with no MPC connectors) to maintain the sterility of tubing sets. In this particular example, vent filters for components 1 and 2 must be removed in a controlled environment to be connected to barbs on an off-the-shelf pre-sterilised sterile filter. This strategy is to avoid re-sterilisation of pre-irradiated off-the-shelf components such as bags and filters. Connection with a standard bag can be different according to the flow rates and bag-system requirements chosen. Design of these sub-systems can be provided for external companies for assembly and/or irradiation. Alternatively, assembly can be done in-house and sterilisation undertaken by autoclaving.

Another consideration during the specification and design of single-use culture bags is in terms of the sensor technology to be used. Some SUB systems have integrated pre-sterilised probes, whereas other systems opt for re-usable optical sensors, or traditional sensors that must be sterilised and re-inserted into the culture bag/vessel. The sensor technology chosen for control and monitoring of cultures may affect batch reproducibility in terms of measurement accuracy, operational flexibility and additional validation of sterilisation. Choice of the correct technology is, therefore, important, and should be considered during the SUB technology trial and as a part of the methodologies and objectives within the wider organisation. Probes that provide greater control of processing and tighter consistency of batch-to-batch products are recommended, particularly for pilot-scale applications in which clinical material is to be manufactured

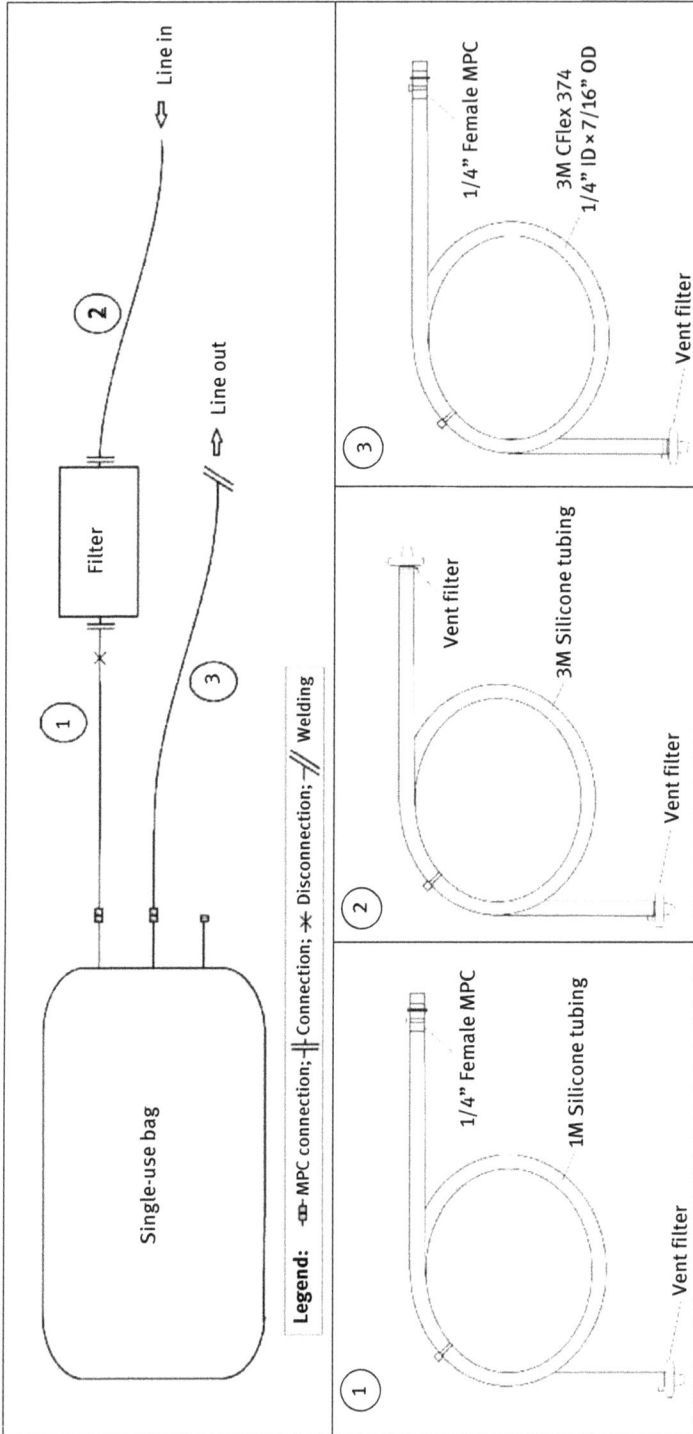

Figure 6.6: Breakdown of the sub-system and individual components.

6.2.3 Risk assessment of the single-use bioreactor process

Table 6.10 presents an example of a FMEA utilised for the risk assessment of a micro-bial culture process using a SUB system. According to the guidelines described in the quality risk management process outlined in International Conference on Harmonisa-tion (ICH) Q9, the first step is identification of the risks associated with failure of processing parameters that must be controlled (and qualified) to ensure reproducibil-ity of the SUB process. In microbial cultures, the critical control parameters are pH, temperature, agitation and DO, which will impact upon cell growth and product yield. Aseptic operation must be maintained because this also impacts upon cell growth and hence productivities. The processing volumes and pressure set out by the bag manufacturer should not be exceeded, to ensure that the integrity of the bag is maintained and connections are kept integral throughout operation, and avoid the risk of contamination. The risks are assessed, and adequate preventive actions and controls to be put in place are suggested. In the case of a SUB operation, these con-cerns are mainly continuous (online or offline) monitoring of process parameters using a validated control system as per requirements set in the URS described in Table 6.7, and operator verification of critical processing stages. The risk rating used to complete Table 6.10 and the criticality of the resulting risk priority number (RPN) can be visualised in the tables presented in Appendix 2. According to these ratings, all SUB operations evaluated that may deviate from targets are categorised as having major potential risks to product quality (RPN between 40 and 216). These ratings are derived from the ability of the system to maintain sterility and control parameters to a set-point, which in turn must be controlled and validated appropriately.

6.2.4 Qualification of single-use bioreactors

Qualification of the culture process in a SUB system should be conducted with the final configuration of the bag and equipment that will be used for a regular manufacturing run. Three different lots of the final design of the culture bag assem-bly, gamma-irradiated to achieve the sterilising dose, should be tested under normal operating conditions (i.e., media, temperature and hold times). Qualification proce-dures should be based upon the requirements established on the URS (Table 6.7) and risk assessment of the SUB operation (Table 6.10). Distinct pass/fail criteria should be established. Based upon these procedures, the integrity, maintenance of sterility, cell growth, API stability and purity should be assessed.

Testing the integrity of final bag assembly, maintenance of sterile connections and sterility should also be assessed under regular manufacturing conditions. For example, maintenance of asepsis and robustness of the culture bag can be mea-sured by filling the culture bag with the maximum volume of media and holding it over a typical culture duration (mammalian culture: ≤21 days; microbial culture:

Table 6.10: FMEA for a SUB operation.

Process step	Failure mode	Effect of failure	S	Potential causes	O	Current controls	D	RPN
pH control	Fail to maintain the pH set-point	Reduced growth Reduced productivities Changes to product profile/quality	6	Controller malfunction Pump failure Calibration failure Cable disconnected from controller Inadequate agitation rates	4	Control and monitoring of pH and agitation rates	2	48
Temperature control	Fail to maintain temperature set-point		6	Controller malfunction Cable disconnected from controller Inadequate agitation rates	2	Control and monitoring of temperature and agitation rates	2	24
DO control	Fail to maintain DO set-point		6	Controller malfunction Calibration failure Cable disconnected from the controller Inadequate air flow or agitation rates	6	Control and monitoring of DO, air flow and agitation rates	2	72
Agitation control	Fail to maintain agitation to set-point level		6	Controller malfunction System disconnected from the controller	4	Control and monitoring of agitation rates	2	48

Pressure control	Fail to measure pressure	Overpressure – weakened connections that result in leaks / Underpressure – low gas flow rates	8	Controller malfunction / Cable disconnected from the controller	4	Visual inspection of culture bag and connections / Control and monitoring of bag pressure	4	128
Cell growth	Lower cell growth	Reduced productivities / Changes to product profile/quality	6	Failure to control process parameters at set-point / Changes in raw materials/starting materials	6	Continuous sampling/monitoring / QC and release of raw materials/starting materials used for processing	4	144
Product yield	Lower product yields	Product loss / Impact upon purification	8	Failure to control process parameters at set-point / Changes in raw materials/starting materials	6	Monitoring of yield and quality level before progress to purification stage / QC and release of raw materials/starting materials used for processing	4	192
Asepsis	Contamination from aeration, media, sampling or leaks	Product loss	8	Failed sterile filter integrity test used for aeration and media sterilisation / Sampling error / Culture bag leaks (culture bag defect, overpressure, tubing welding fails)	4	Filter integrity pre- and post-use/operation / SOP with defined sampling procedure / Visual inspection of bag integrity pre- and post-assembly / Control and monitoring of bag pressure / Maintenance of validation of ancillary equipment used for welding the tubing sets	6	192

DO: Dissolved oxygen
QC: Quality control
SOP: Standard operating procedures

≤14 days). This test should be undertaken at normal operating temperature as well as a maximum agitation and a gas-flow rate that results in over-pressurisation of the bag [19]. Depending upon the risk assessment, samples of solution can be taken to assess sterility, measure the impact of particulates generated during mixing, and substances that may leach from the plastic film that can interfere with cell growth or with the purity of the API contained within the bag. After qualification of the SUB system, there should be continuous monitoring of the significant production parameters identified, as well as other critical attributes that may affect the quality and yield of the final API product. The process should be re-qualified at regular intervals. Additional details of qualification of final bag assemblies and culture bag systems can be obtained from Section 6.1.

6.3 Case study 3: Tangential-flow filtration

Ultrafiltration (UF) is the most widely used form of TFF. UF is used to concentrate small biomolecules while removing impurities and contaminants from solution. Membrane filters are used to retain biomolecules but allow buffer through their porous structure into a separate permeate stream. Diafiltration (DF) is a TFF process in which buffer is introduced into the recycled retentate tank while the filtrate is removed. TFF in combination with DF are used to exchange buffers and reduce the concentration of product or undesired species. Microfiltration (MF) membranes can also be used in TFF systems. The larger pore size of MF membranes means that they are applied to different types of feed streams, and can be used to harvest and wash cell suspensions, or clarify cultures and cell lysates.

TFF flat-sheet cassettes or hollow-fibre cartridges have existed in single-use format for many decades. However, typically these cassettes would be integrated within stainless-steel skids that would supply the fluids to the cassette alongside measurement and control of process parameters. In recent years, fully integrated skids combing single-use flow-paths with cassettes have entered the market at a range of scales covering the laboratory bench through to pilot-scale manufacturing, and with varying levels of automation and control of processes. These systems accommodate various TFF cassettes with different membrane types and sizes, as well as different capacities of re-circulation tanks to support higher processing volumes and/or concentration factors. However, these platforms have limitations if high flow rates are required for systems with high membrane areas. In general, TFF cassettes have similar performance and scalability to that of re-usable products at a fraction of the cost. These cassettes can be supplied pre-sterilised and pre-sanitised, ready to be equilibrated with buffer and used for processing. These factors provide greater flexibility and reduction of costs associated with product change-over in multi-product processing because product contact parts can be discarded after use, which reduces the risk of cross-contamination between batches.

6.3.1 Selection of technology for tangential-flow filtration

Selection of a suitable TFF system starts with definition of the purpose of operation and assessment of the volume and concentration of the material to be processed, alongside the conditions of the product at the end of the operation. Then, the type of membrane filter, surface area, pore size, required membrane flux and cross flow rate require consideration to suit the application. These conditions could be based on a scale-up from a smaller process. However, these conditions also may not have been determined, and the system may need to be flexible to meet the requirements of future processes and products. Demonstration that the TFF cassette has at least equivalent performance to a re-usable system can provide confidence regarding the scalability of the system [20]. Such demonstration can be achieved by use of small-scale systems and evaluation of parameters such as cross-flow rates, transmembrane pressure (TMP), processing time, recoveries, and membrane flux.

The mode of operation of the system should also be considered. Each mode of operation will have different set-points that would be required for the system if operated at, for example: constant TMP or flux; in concentration, continuous or discontinuous DF, feed vessel filling; during normalised testing of water permeability and flushing.

Another key consideration when specifying or selecting a TFF system is the integrity and compatibility of materials of construction with the process fluids to be used. This is done to avoid conditions whereby the properties of the single-use system may be degraded to the point where they no longer function reliably or lose integrity, and to ensure that the single-use system does not leach toxic compounds or excess particulates are released into the API product stream (see Section 6.1 for more details on considerations for extractables and leachables).

During selection of the single-use TFF system, the end-user must decide: the scale of operation; whether a user-defined or off-the-shelf system will be used; the required design features of the system; the vendor capabilities to meet the requirements of the process and business. These considerations can be summarised in a strength, weakness, opportunities and threats analysis (SWOT) analysis to visualise strengths and weaknesses of the single-use TFF system under consideration, an example of which is given in Table 6.11. Smaller-scale applications are more readily amenable with user-defined customised single-use flow-paths because design features are simple, the number of components required to be assembled is relatively small, and control of system can be achieved by use of a combination of single-use sensors and standard laboratory equipment. Single-use sensors are available for measurement of process parameters such as pressure or flow rate that can, in turn, be connected to balances and data-loggers to recover data from small-scale processes. As the scale increases, the complexity also increases, with the number of components requiring set-up and testing pre- and post-use augmented, and an increased risk of failure and errors. If the supply of clinical material is required, then greater batch-to-batch consistency is required. To meet these requirements, off-the-shelf

Table 6.11: SWOT analysis of single-use TFF.

Strengths	Weaknesses
Eliminate CIP and cleaning validation	Scale of operation. Most beneficial at small-scales (<5 m2) and if the frequency of cassette change-over is high
Reduce risk of cross-contamination between batches and campaigns	Increase in the cost of consumables (single-use cassettes and flow-paths)
Decrease the number of process steps by reducing processing time and labour requirements	Complexity of installation and high number of disposable flow-path components
Off-the-shelf systems are available Sterile and pre-sanitised disposable flow-path ready to be used	Additional equipment may be required (welder, sealer)
Reduced footprint	Limitations in pump capacity and tubing sizes that are a part of re-circulating systems to handle high flow rates and pressures
Scalable system Automatic process control	Loss of product due to large system hold-up volumes
Automatic monitoring and acquisition of data	Limitations in terms of maximum processing volume or filter area/type capacity
Broad range of applications (MF, UF)	

Opportunities	Threats
Increase process efficiency by elimination of non-value added steps	Vendor technical support
Short equipment turnaround times	Vendor can supply numbers of consumables of appropriate quality
Performance consistency of single-use cassettes	Lack of inter-changeability Immature system
	System integrity
Adds flexibility in a multi-product facility	Leachables and extractables from filter cartridge and disposable flow-paths, particularly if system is used for the final formulation
Reduced capital costs, faster construction and installation	Biocompatibility
	Suitable small-scale model
	System design may be new and untested for clinical or commercial applications

single-use TFF systems provide greater control of process parameters, with reduced operator interaction. However, they are off-the-shelf systems with pre-set designs and functionalities, so they have limited capabilities as well as set capacities. Even though a broad range of membrane formats and equipment used in single-use TFF systems are available, the design may not always be suitable for the intended application. In these instances, a compromise may have to be made between technical assessment and business assessment. The maturity of the system also plays an important part in the decision to select a particular single-use TFF technology. If the single-use TFF system is in the prototype stage, the end-user may be able to develop the system in collaboration with the supplier and influence the design of the system for the intended use. This strategy provides the opportunity for the end-user to familiarise personnel with novel technology while testing the system. Alternatively, if a TFF system is mature and application has been proven with a wide range of process streams/products, this scenario may assure the end-user that the system has undergone many design cycles that have improved its functionality (e.g., reduced hold-up volumes, customised leak-proof flow-path assemblies, defined protocols). Finally, if these systems are not compatible with other single-use systems and equipment available at the end-user site, this limits the end-user in case the supply of consumables is interrupted or quality issues arise from supplied items.

6.3.2 Specification and design

The URS for a TFF system is similar in layout and content to the one shown in Section 6.2 for a SUB system. An example of an URS for an off-the-shelf integrated TFF skid is given in Table 6.12. The main difference to the SUB URS is the requirements of process parameters such as membrane area/cassette capacity, flow rates and fluxes, TMP, location and requirement of pressure indicators and flow meters. In addition, the requirement of the single-use flow-path assemblies that are in direct contact with API are included to enable the supplier to verify that the design is suitable for the intended application and to ensure that the equipment is qualified appropriately. Cleanliness of materials and compatibility with the API and cleaning solutions are also important specifications. They are specific to a process and API, so should be based upon a risk assessment.

For a TFF system, it is likely that several solutions will pass through the system boundaries. For manufacturing applications, it is necessary to understand what these are so that the design of the system has sufficient inlets and outlets to handle all of the solutions. For example, in most TFF designs, it is necessary to have an inlet manifold that allows for multiple bags to be pre-connected to the system or to contain a sufficient number of inlet connections to allow bags to be connected to the system throughout the operation. Flow diagrams are a useful tool to capture and review this information. A block flow diagram, such as that shown in Figure 6.7, can be used to summarise the flow in and out of the process based upon the unit operations that take place within the process. Ideally, the volumes of solutions entering and leaving the system would be known. However, in the early stages of process design, the volumes may not be fixed, so ranges that the system might have to cover should be considered instead. It can be seen from Figure 6.7 that several solutions enter and exit the system, from product streams, buffers, compressed air and samples. Solutions enter and leave throughout the course of the operation, so they must already be pre-

Table 6.12: Single-use TFF and URS.

Operational requirements	
Capacity	The system must allow for UF and DF product stream of API mixture within a buffer system. This includes the concentration of the feed from 'X' to 'Y' l by a concentration factor of 'X', and the DF at 'X' l for 'X' diavolumes
Process requirements	Compatibility with membrane cartridge type and filter area Target filtration time TMP Cross-flow rate Permeate rate Temperature control Concentration factor Automatic dosing of feed tank Leak tested before use Hold-up volume Feed pressure
Process control parameters	Control parameters, measurement range, accuracy/ tolerance and duration: – Temperature of feed vessel – Agitator speed – Cross flow rate – Permeate flow rate – Feed tank weight – Permeate tank weight – Feed pressure – Retentate pressure – Permeate pressure – Conductivity – Volumetric concentration factor

Control system
Comply with 21 CFR Part 11 and built in accordance with GAMP® principles Shall be able to store sufficient amounts of data Allow configuration of all equipment parameters The control system will be the means of monitoring and controlling the equipment Viewing and resetting alarms Print batch information Able to configure all equipment parameters/recipes

Data and reporting	All parameters will be stored, trended and will be available for viewing and printing Control system shall be able to generate a batch report

Table 6.12 (continued)

Alarms and safety features	Power failure recovery Emergency stop Interlocks to prevent system overpressure Alarms and warnings (see example of **Table 6.9** for a list of critical alarms applied to a SUB) Information messages Data and security levels (see example of **Table 6.8** for security levels applied to a SUB) Data collection
Interface requirements	Electrical supply type Filter holder Sensors Weldable lines Bag connections Sample points
Process contact fluids	The system contact materials should be compatible with the following fluids: WFI, sanitisation solution, DF buffer, product contained in buffer, equilibration buffer
Hold-up volume	Minimised and specified by vendor
Sampling	Samples should be able to be taken from the permeate and retentate lines
Functions	The system should operate with minimal involvement of the operator The system shall be operated locally by means of an operator interface terminal
Materials of construction	The support frame and associated equipment should be suitable for cleaning to have sufficient access for cleaning Materials of construction should be suitable for ISO 8 cleanroom environment Suitable single-use assembly materials of construction compatible with process fluids that maintain the integrity of the product and operations. Single-use assembly (see requirements in Single-use Assemblies section below for more information)
Utilities	The operation of the system must be able to function with the available utilities: – Electricity – Process gases (O2, N2, CO2) at 'X' barg – Process 'X'% gas/air mix at 'X' barg – Clean steam and non-clean steam at 'X' barg – PW at nominally 'X' barg – Chilled water supply at 'X' barg and 'X' C – Chilled water return at 'X' barg – WFI at nominally 'X' barg – Towns water at 'X' barg – Instrument compressed air at 'X' barg – Clean compressed air at 'X' barg – Biowaste drain

Table 6.12 (continued)

Environment	Layout (attach drawings with dimensions of room and equipment) Room classification Cleaning requirements
Quality and regulatory requirements	CE certification Factory acceptance test Site acceptance test protocol (and test upon completion) Installation, operation and maintenance instructions (and full report upon completion) Instrumentation and control systems should comply with: – 21 CFR Part 11 – Electronic records, electronic signatures – GAMP® 5 – Guide for validation of automated systems – EU Annex 11 Volume 4 All single-use assemblies should, as a minimum, comply with the following: USP Class VI Biological test for plastics ISO 10993 Biological evaluation of medical devices ISO 11137 Sterilisation of healthcare products Single-use TFF filter cassette should be supplied 100% integrity tested
Requirements for single-use assemblies	See Figure 6.9 for retentate bag design and tubing lines Retentate bag requires mixing Assembly of single-use manifolds and tubing sets should take place inside of Cleanroom grade D (ISO 8) Single-use assemblies that have direct product contact should have as a minimum: – USP Class VI Biological test for plastics – ISO 10993 or USP 87/88 – Biological evaluation of medical devices – ISO 11137 Sterilisation of healthcare products – Assurance of sterility (gamma irradiation) – TSE/BSE free manufacture (EP 5.2.8) – EP 2.6.14 or USP 85 – bacterial endotoxins test – USP 643 – Total organic carbon

connected to the system during set-up, or the system must have a provision by which the solutions can be connected and disconnected to the system during operation. A further consideration is that, if the scale is large and buffer volume requirements high, then there may be multiple hold bags of a given buffer that will require connecting to the system during a given process step within the operation.

A PFD, such as that illustrated in Figure 6.8, provides further detail to the design of the operation. The PFD is not as detailed as a piping and instrumentation diagram (P&ID), but does show additional information, such as:
- How solutions connect to the system;
- How waste and the product leave the operation;
- The boundaries of the system;
- Simple overview of how single–use components interact with equipment hardware; and

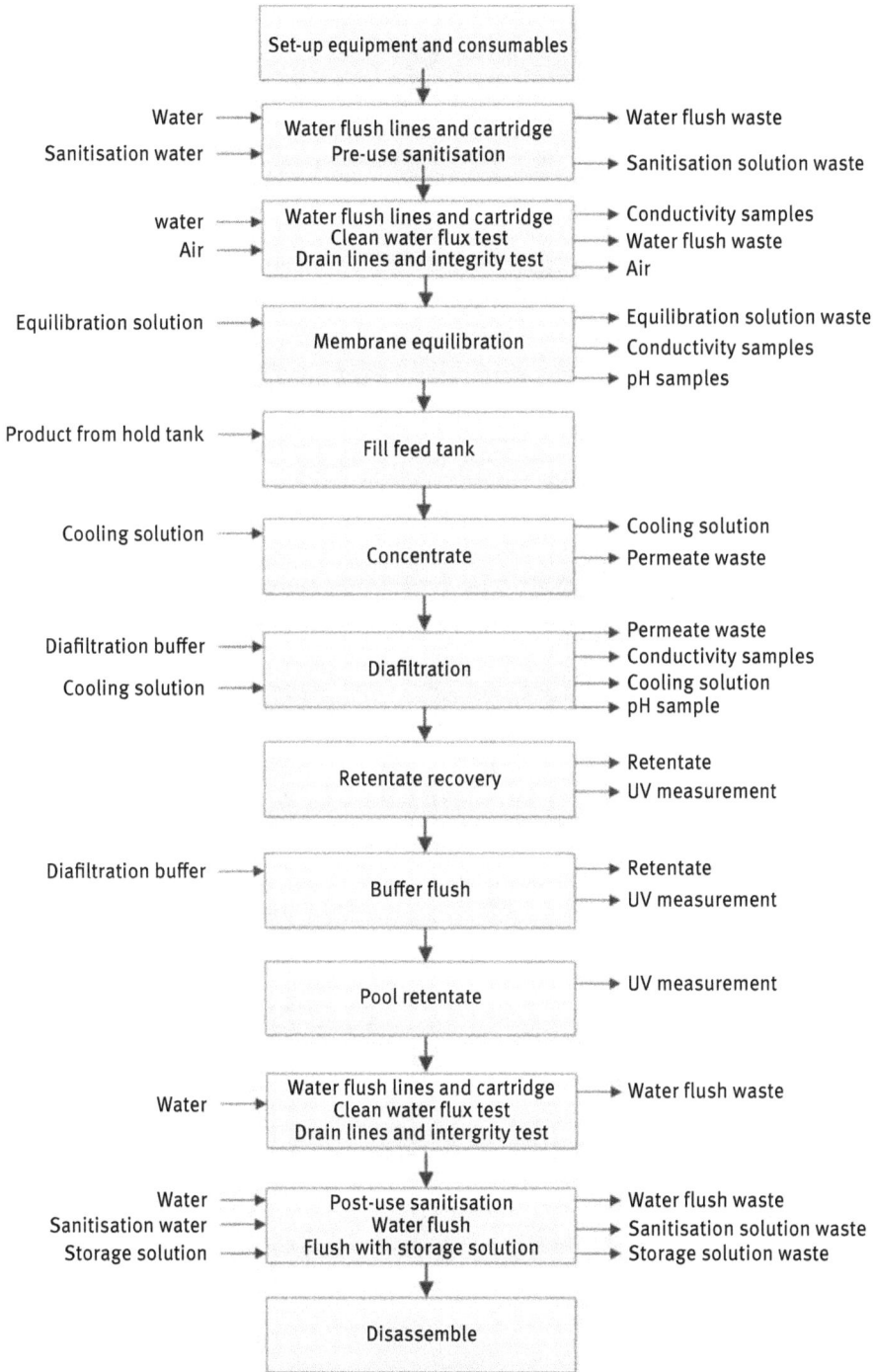

Figure 6.7: Block flow diagram for TFF operation.

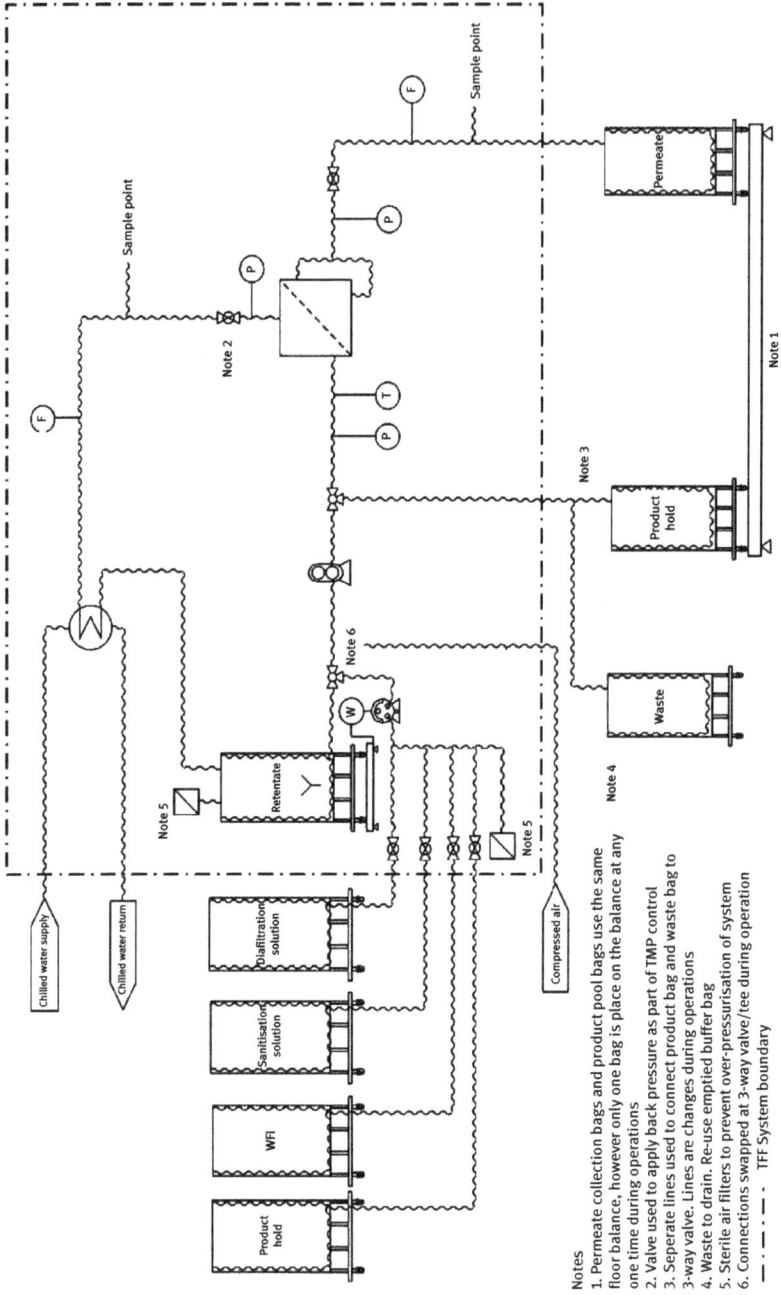

Notes
1. Permeate collection bags and product pool bags use the same floor balance, however only one bag is place on the balance at any one time during operations
2. Valve used to apply back pressure as part of TMP control
3. Seperate lines used to connect product bag and waste bag to 3-way valve. Lines are changes during operations
4. Waste to drain. Re-use emptied buffer bag
5. Sterile air filters to prevent over-pressurisation of system
6. Connections swapped at 3-way valve/tee during operation
—·—·— TFF System boundary

Figure 6.8: PFD of TFF operation.

- Some of the key elements of measurement and control to be used within the system.

For the TFF design shown in Figure 6.8, several measurement and control elements are required:
- Pressure on the feed, retentate and permeate lines to control TMP;
- Temperature measurement, which could be used to control the degree of cooling achieved with the heat exchanger on the retentate line; and
- Flow in the retentate and permeate lines can be used to control concentration and DF operations; alternatively, the weight measurement from the permeate side balance can be fed into the control system.

Standard single-use components are available from different suppliers to measure pressure, temperature and flow. However, the feasibility of integration into the control system must be considered, alongside the accuracy of the measurement over the course of the operation. If a control system is not utilised, then manual intervention by the operator is required to set the cross-flow rate and retentate back-pressure to achieve a given TMP and/or permeate flow rate. In addition, the weight of the permeate bag would require monitoring to ensure that the correct volume of buffer solutions are collected to achieve the given product concentration and DF flushes.

The tubing sets required can be broken down into key sub-systems (Figure 6.9):
- *Feed manifold line*: An inlet manifold to provide solutions into the system *via* a pump into the three-way valve located in feed tank to feed pump line. The pump is controlled *via* the weight of the retentate bag to ensure that the correct volume of solution is dosed into the bag. A hydrophobic vent filter is used to prevent build-up of air in the lines, or over-pressurisation;
- *Feed tank to feed pump line*: A tubing line connecting the feed/retentate bag to the inlet to the positive displacement pump, and providing a means for the feed manifold line to be connected to the TFF system;
- *Feed pump to filter line*: A tubing line connecting the outlet from the feed pump to the inlet of the TFF filter housing. In addition, a three-way valve provides a means to connect the line to bags to collect waste solution or retentate;
- *Product line*: A product line connecting to the feed pump to filter line *via* the three-way valve so that the product can be pumped from the retentate bag;
- *Permeate line*: A tubing line connecting to the permeate side of the filter housing so that the waste buffer solution can be collected and disposed of;
- *Retentate line*: A tubing line connecting the feed side outlet of the filter housing back to the retentate bag, and passing through a heat exchanger to control the temperature of the solution containing the product; and
- *Feed tank*: A feed/retentate bag with inlets, outlets, a hydrophobic sterile vent filter and mixing system to maintain a homogeneous solution.

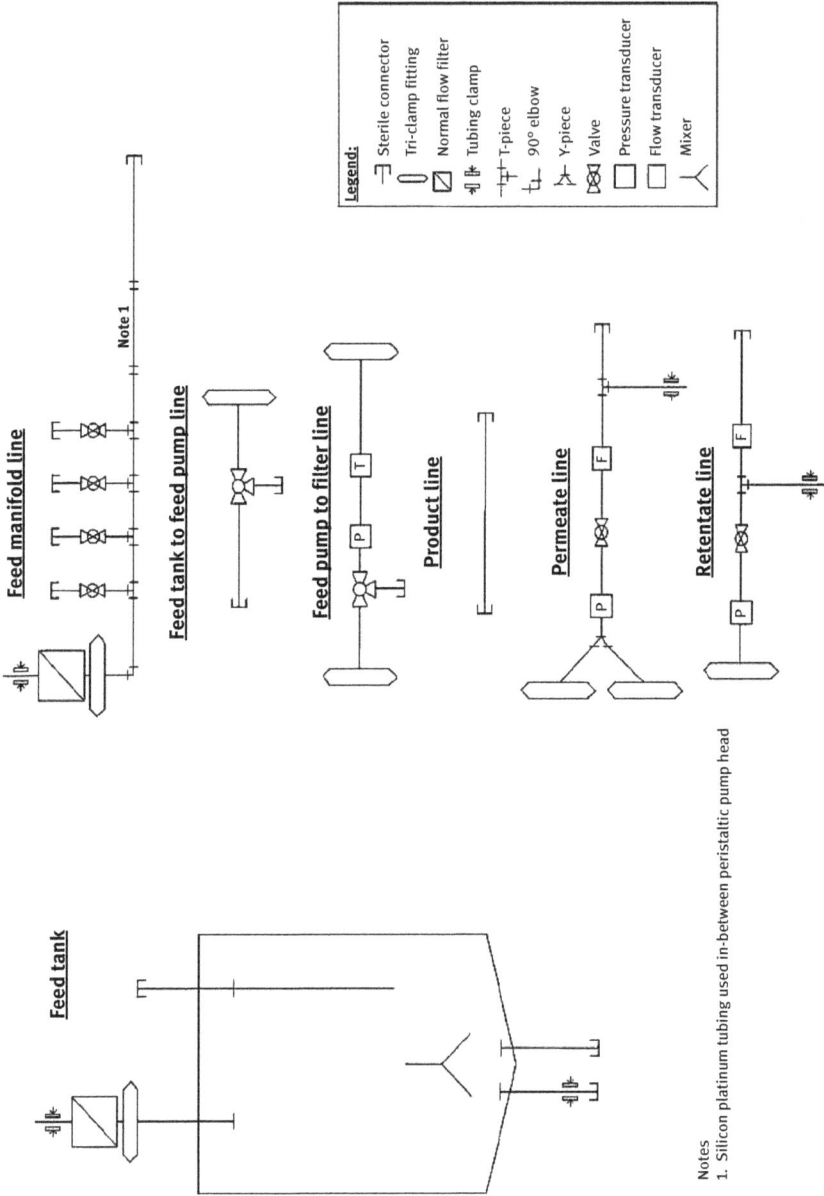

Figure 6.9: A single-use sub-system for TFF operation.

The PFD provides a sufficient level of detail if a supplier-designed system is to be utilised because it portrays the intent of the system, and can be coupled to a URS as part of a package of specification documents to be passed to potential suppliers for review. If a simpler system is required and is to be designed and built in-house, then further details of the single-use sub-system are required so that they can be fabricated in-house or through a third-party supplier. Examples of such drawings are given in Figure 6.9 for the sub-systems described above. In these drawings, greater detail is given to the individual components that make up the sub-system. If multiple single-use systems are being developed, then these components can be compiled during the design phase for review with the aim to develop standard components for tubing and aseptic or septic connections (see Section 6.1 for approaches to standardisation of bag assemblies).

These diagrams will probably undergo numerous changes throughout the design process and, as such, they should be version-controlled. Once finalised, the drawings can be used to review the built system when conducting site acceptance testing (SAT). They are also useful tools to familiarise an operator with the equipment, consumables and operations.

6.3.3 Risk assessment to support design of a system

The TFF system comprises several separate single-use sub-systems, and each must be identified and the design assessed for any weaknesses or failure points. Complexity of the TFF operations can be seen from the PFD (Figure 6.8) and single-use system drawings. Due to this complexity, it is advisable to consider the risk of failure of the system at an early stage of the design process, when changes can be made, or the approach can be rejected and another considered. At an early stage of the design and with a lack of hands-on operational experience with the system, a simpler preliminary risk assessment (PRA) is preferable to a more complex FMEA approach. The PRA can be used to identify potential failure risks which can, in turn, be used to prioritise review of the design, conduct initial testing of the system before purchase, or engage with suppliers to identify if they have data to show that the risk can be mitigated. The approach is applicable to off-the-shelf or customised designs.

Figure 6.8 is a useful tool to assist with the PRA because it shows the interaction between single-use components and equipment hardware, and the interaction between different single-use systems with each other. It also helps visualise the boundaries of the overall TFF system and how it interacts with other operations. Figure 6.9 can also be used as part of the PRA because it shows the discrete single-use sub-systems and components that make up the TFF system and the connectivity across sub-systems. An example of a PRA applied to a single-use TFF system is given in Table 6.13 for two of the single-use sub-systems: TFF feed tank and feed pump to a TFF filter line.

Table 6.13: Preliminary risk-assessment of TFF sub-systems: feed tank and tubing line from pump outlet to filter inlet.

Purpose of system: feed tank

Assembled by two operators, by inserting into TFF feed tank holder. Inflated before operations by use of compressed air connected *via* interface with process air within the cleanroom.

Holds a range of process fluids during TFF operation:

– Product feed from previous unit operation
– Concentrated product
– Diafiltered product
– Cleaning solution
– Water
– Air

Provides mixing to process fluids to ensure homogenous solution.

Feed tank is filled automatically from a feed manifold *via* interface with a feed tank to feed the pump line. Fill volume is set by the operator and controlled *via* a balance upon which feed tank holder is on top of.

During operation, liquid is removed from feed tank *via* interface with the feed tank to the feed pump line.

During draining, process fluids are removed via the drain line. When not in use, this line is clamped shut.

Process fluids return to the feed tank after TFF operations *via* interface with the retentate line.

Over-pressurisation within the feed tank is prevented by the air filter attached to the top of the tank.

Name	Potential failure mode	Potential effects	Causes	Action
Feed bag	Loss of integrity	Loss of product	Seals of bag film damaged during fabrication or transportation	Review transportation methods
		Exposure of operator to process fluids		Visual inspection before use
				Integrity test before use
		Termination of operation		Monitor quality across batches
		Contamination	Bag components degrade due to gamma irradiation	Review supplier tests
				Conduct in-house 100% leak test with systems gamma irradiated with sterilising dose
			Assembly by operator	Undertake tests on ease of use of assembly with operators during the design phase and before selection
				Review operator procedures and training
			Over-pressurisation during leak test	Carry out in-depth review of root causes, mitigation methods and probability of occurrence
			Incompatibility with process fluids	Review supplier data of application of single-use systems with materials
				Conduct process tests with components/fluids of concern
			Tubing welds fail	During design:
				− Review supplier tests and any data on number of failures
				− Test systems under pressure to detect failure point Post-design:
				− Integrity test before use
				− Monitor failures across batches

(continued)

Table 6.13 (continued)

Name	Potential failure mode	Potential effects	Causes	Action
			Drain clamp fails	During design: – Review supplier tests and data on number of failures – Test systems under pressure to detect failure point – Consider double clamp or secondary isolation system Post-design: – Integrity test before use – Monitor failures across batches
			Drain clamp left open	Review operator procedures and training
			Failure of connection with retentate line	Test robustness of connection
			Failure of connection with feed tank to feed pump line	Review operator procedures and training
			Too much mixing	Test mixing performance at the top end of the mixing rate
			Not enough mixing	
			Failure of interface with mixing hardware	Review supplier data for failure of the system Conduct tests of mixing at high mixing rate across long operational time periods using a viscous solution

Air filter	Does not allow sufficient gas flow	Blocked due to contact with liquid	Conduct root-cause analysis to identify potential causes
			Review design to assess likelihood of occurrence
	Bag over-pressurisation leading to failure	Installed incorrectly during fabrication	Test ease of assembly with operators before final selection
			Review operator procedures and training

Purpose of system: feed pump to filter line

Connect to the outlet of the pump-positive displacement pump head.

Connect to the filter holder inlet.

Measure pressure of fluid entering filter holder.

Measures temperature of the fluid entering the filter holder.

Connects to bag systems used to collect waste or product.

Three-way valve to control direction of process fluids to the filter holder or bags to collect waste or product.

Exposed to a range of process fluids during TFF operation:

– Product feed from previous unit operation
– Concentrated product
– Diafiltered product
– Cleaning solution
– Water
– Air

(continued)

Table 6.13 (continued)

Name	Potential failure mode	Potential effects	Causes	Action
Tubing line	Leaks	Loss of product Exposure of operator to process fluids	Tubing fails immediately	Review robustness of tubing and connectors post- gamma irradiation at sterilising dose Review supplier tests
		Termination of operation Contamination	Tubing degrades over operational time, then fails	Review robustness of tubing and connectors after exposure to worse-case process fluid for long operational periods Review supplier tests
			Tubing connectors fails	Test robustness of connection Review operator procedures and training Review supplier tests
			Tubing set is not well assembled by the supplier	During design: – Review supplier tests – Review sample set of tubing Post-design and selection: – Monitor supplier performance – Visual inspection
			Incorrect assembly of tubing within equipment	Test ease of assembly with operators before final selection Review operator procedures and training
			Incompatibility with process fluids	Review supplier data of application of single-use systems with materials Conduct process tests with components/fluids of concern

Pressure sensor	Leaks	Loss of product	Component failure	Review supplier tests
		Exposure to operator to process fluids		Conduct in-house test with systems gamma irradiated with sterilising dose
		Termination of operation	Connection with tubing fails	Test robustness of connection
				Review operator procedures and training
		Incorrect pressure measurement	Component incorrectly assembled	Test ease of assembly with operators before final selection
		Contamination		Review operator procedures and training
			Component degrades over operational time, then fails	Review robustness component after exposure to worse-case process fluid for long operational periods
			Incompatibility with process fluids	Review supplier data of application of single-use systems with materials
				Conduct process tests with any components/fluids of concern
Pressure sensor	Incorrect pressure measurement	Incorrect TMP with potential impact upon filtration operation	Incorrect calibration	Test sample of sensors to ensure consistency across lots
			Component failure	Review supplier tests
		Over-pressurisation leads to product degradation		Conduct in-house test with systems gamma irradiated with sterilising dose
			Incorrect assembly	Test ease of assembly with operators before final selection
				Review operator procedures and training
			Shift in measurement accuracy during operation	Monitor performance with worse-case conditions over long operational time
			Sensor not in contact with process fluid	Review design and requirements of sensor
			Leak from unit	Test component under worse-case process conditions over long operational times

(continued)

Table 6.13 (continued)

Name	Potential failure mode	Potential effects	Causes	Action
Temperature sensor	Leaks	Loss of product	Component failure	Review supplier tests
				Conduct in-house test with systems gamma irradiated with sterilising dose
		Exposure of operator to process fluids	Connection with tubing fails	Test robustness of connection
				Review operator procedures and training
		Termination of operation	Component incorrectly assembled	Test ease of assembly with operators before final selection
		Incorrect temperature measurement		Review operator procedures and training
			Component degrades over operational time, then fails	Review supplier tests
				Conduct in-house test with systems gamma irradiated with sterilising dose using worse-case process conditions over long operational period
			Incompatibility with process fluids	Review supplier data of application of single-use systems with materials
				Conduct process tests with any components/fluids of concern
	Incorrect temperature measurement	Poor temperature control, leads to exposure of product to high temperature and results in product degradation	Incorrect calibration	Test sample of sensors to ensure consistency across lots
			Component failure	Review supplier tests
				Conduct in-house test with systems gamma irradiated with sterilising dose

Poor temperature control, leads to exposure of product to high temperature and results in product degradation			Incorrect assembly	Test ease of assembly with operators before final selection Review operator procedures and training
			Shift in measurement accuracy during operation	Monitor performance with worse-case conditions over long operational time
			Leak from unit	Monitor for leaks with worse-case conditions over long operational time
Interface with collection bags lines	Leaks	Loss of product	Component failure	Review supplier tests Test robustness of connection Conduct in-house test with systems gamma irradiated with sterilising dose
		Exposure of operator to process fluids	Incorrect set-up by operator	Test ease of assembly with operators before final selection Review operator procedures and training
			Incompatibility with process fluids	Review supplier data of application of single-use systems with materials Conduct process tests with any components/fluids of concern
		Product collected in waste bag	Incorrect set-up by operator	Test ease of assembly with operators before final selection Review operator procedures and training
Incorrect bag attached during operation		Waste collected in product bag		

The PRA is a 'brainstorm' of potential failure points. It identifies actions for further investigation during the design stage. The risk assessment should be updated during the design verification. If testing has been conducted with actual systems, then a more complex assessment such as FMEA (see examples of FMEA for SUB in Section 6.2 and fill-finish in Section 6.4) can be used, whereby the probability of failure is assessed and quantified.

6.4 Case study 4: Formulation and fill-finish

Operations in an aseptic fill-finish line comprise various activities, such as formulation, filtration, solution transfer and dose forming. Adoption of SUT within this process improves the flexibility, reliability and efficiency of filling and associated processes. Flexible filling lines (in which the entire product path is single-use) contribute to rapid change-over between different products. Cleaning and SIP of filling pumps and transfer piping is replaced with assembly of an un-used pre-sterilised tubing set and product bag. Single-use assemblies can be combined with mixing systems used for formulation and various dosing systems for filling the drug product into various container types (vials, syringes, ampoules, bags). This approach provides flexibility, particularly for multi-product facilities. Provision of a completely sterile product flow-path (closed system) and the disposable nature of the operation reduces the risk of cross-contamination between batches of different products.

SUT are available for application throughout the formulation and filling process. Mixing container systems used during formulation can use different mixing technologies for powders and liquids, within a closed or open environment, with different configurations that are customisable with a wide range of fittings and tubing. Sterilising-grade filtration systems are available widely. Transfer technologies include weldable tubing and sterile connectors.

The choice of pump technology in filling and dosing systems is very important. Many technologies are available from peristaltic to time-pressure filling systems, along with use of positive-displacement, diaphragm or rotary pumps. Advantages and disadvantages of different types of dosing technology are based upon application, complexity, speed, scale and potential effect upon the drug product. Rotary piston pumps (and their variant, diaphragm pumps) are the most commonly used equipment for filling due to their wide range of viscosities and temperatures. Use of these pumps with shear-sensitive material can, however, be an issue. Peristaltic pumps can also be used but have a limited viscosity range and lower precision than rotary pumps. Time–pressure systems are more accurate than the other pump systems mentioned and create very low shear in the product. These dispensing systems are then used for filling containers such as vials, syringes, ampoules, and bags. The properties of the product, storage, dosage and method of administration are crucial for choosing the most appropriate final container to be used.

6.4.1 Selection of fill-finish technology

Most fill-finish operations require various components such as tanks for liquid mixing, piping for transfer, and ancillary equipment such as a pump or filtration apparatus. Figure 6.10 shows the filling operation and possible design configuration of a disposable manifold system for products to be filled into vials using a filling needle or aseptically into bags. This design shows a formulation step followed by an in-line sterile filtration and a hold step in which the sterile formulation is contained within a filling bag immediately before the dosing-pump or filling machine. Once the integrity of the sterile filter is confirmed, the sterile formulation can be aliquoted into the final containers. This schematic representation also shows potential cleanroom classifications for each stage of the process and single-use component/equipment location. Assuming the SUT filling assembly is a contained closed system, non-sterile and sterile formulations can be undertaken and held in a cleanroom class-C area. Given the increased risk of contamination of open manipulations using the filling needle, dispensing into vials has been placed in a class-A cleanroom environment. If dispensing is undertaken using a closed system with aseptic disconnection such as the one shown for filling bags, this operation could be placed in cleanroom class C according to the risk associated with this activity. Choice of the design of the single-use fill-finish configuration should be based on minimising product losses due to in-line hold-ups and filter integrity tested after use (use of aseptic disconnection technology). This choice can be achieved by careful mapping of fill processes and exact tubing lengths, followed by small changes in the fill procedure and choice of disposable systems (connectors, filters). During the filling operations of a sterile product, the volume, filling rate, accuracy and sterility maintenance, as well as the level of automation, are important factors to consider.

Many suppliers provide pre-assembled and sterilised filling assembly designs. These should be tested thoroughly before purchase. These tests can be done with water to evaluate the integrity of the system and to see if process conditions are compatible with SUT materials. A 100% leak test pre- and post-operation, wear and tear of silicone tubing during pumping action, and generation of worse-case- scenario particulate and leachable profiles are examples of parameters that should also be evaluated during a trial. Irrespective of whether the end-user chooses an off-the-shelf system or custom-made design, product development teams should work closely with SUT suppliers as well as manufacturers of the final container and pump system to assess the impact of the chosen production filling system upon the drug product.

6.4.2 Risk assessment of fill-finish

During production of a sterile drug, aseptic formulation and filling of the final product are the most difficult processes to qualify due to the high potential impact upon

Class C area

Class A area

Filling needle

Filling bag

Aseptically connected bags

Waste bag

Vent bag

Mixing bag

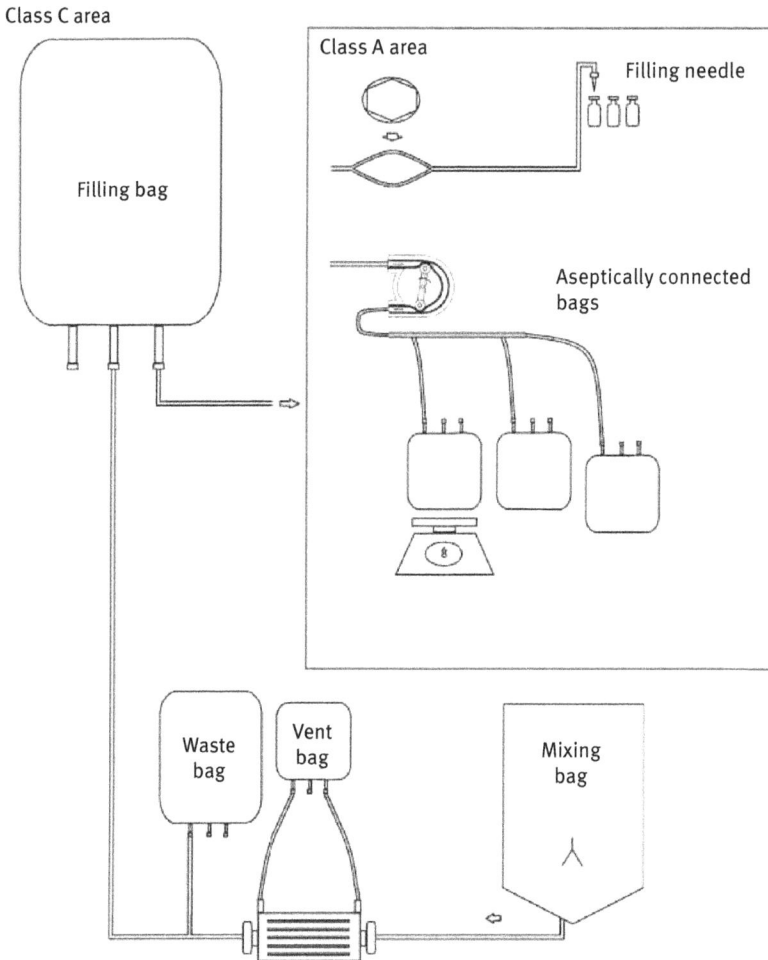

Figure 6.10: Detailed SUT fill-finish approaches. Class-C area: formulation followed by sterile filtration into a filling bag. Class- A area: dosing pump and filling needle used to fill vials; aseptic connection from the pump into several bags - product suspension is dispensed into bags to a pre-set final weight.

the patient. Contamination is a high risk for a fill-finish operation. Contaminants can be introduced by personnel, airborne dispersion, or through poor sanitisation or sterilisation of equipment and compounds. Potential causes of contamination can be visualised readily in a Fishbone diagram (Figure 6.11), whereby the cause and effect of non-conformances derived from materials, controls, personnel, equipment, and procedures can be analysed. In a fill-finish operation, reduction of the risk of contamination requires careful control of the aseptic environment *via* extensive monitoring of the environment, personnel practices and procedures, and sterilisation of equipment and components.

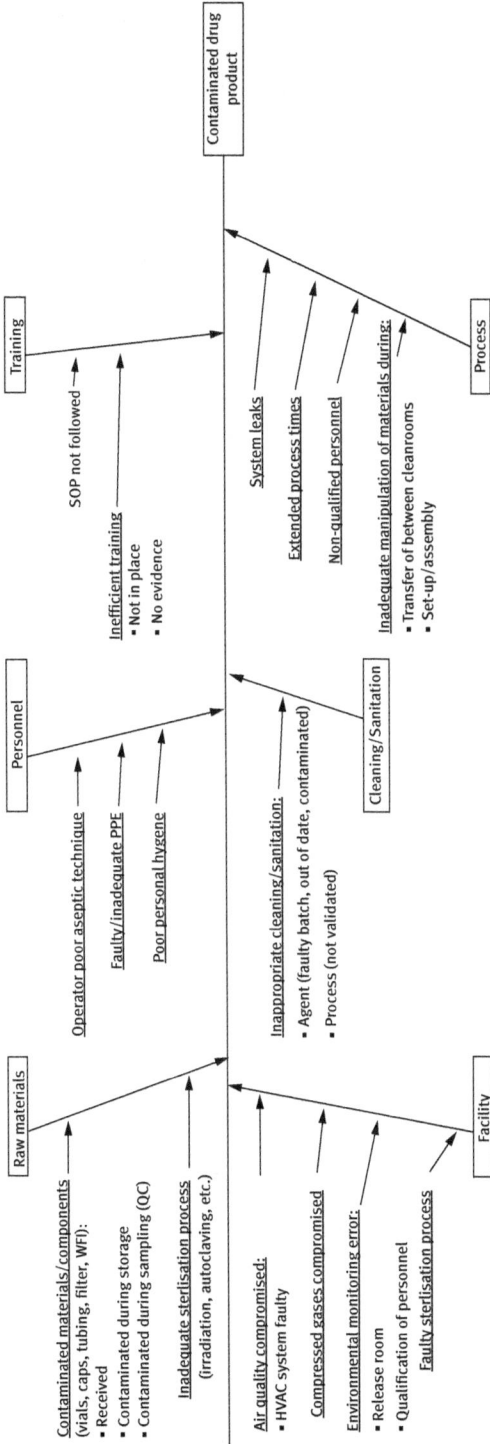

Figure 6.11: Fishbone diagram showing potential causes of contaminated product. HVAC: Heating ventilation air conditioning; and PPE: personal protective equipment.

Another risk during a fill-finish operation is process variations or introduction of substances that may affect the purity and/or stability of the final drug product. Process variations can affect the dose form through insufficient homogeneity of the dispensed suspension, leading to an incorrect quantity of product in the final container (vials, syringes, bags). Substances such as particulates and leachables derived from filters, plastic material or filling needles may interfere with the stability and purity of the drug product. The fill-finish operation is at the end of the production process, so substances introduced into the system are unlikely to be removed before administration to the patient. Hence, they must be identified and quantified, and the impact upon product quality and patient safety assessed. One of the proposed tools in ICH Q9 guidelines that can be used to examine the impact of potential failures upon product quality, and to propose adequate corrective and preventive actions, is by undertaking a FMEA (Table 6.14). A FMEA helps to establish control strategies to eliminate or reduce the risk to an acceptable level. In the case of a fill-finish operation and in accordance with cause-and-effect analysis (Figure 6.11), post-testing of cleanrooms and sterile materials as well as use of trained personnel ensures that the risk of contamination is reduced. Similarly, calibration of the filling system and continuous verification of parameters (e.g., mixing speed, time and filling weights) ensures that the drug product is homogeneous and that the right dose is dispensed. Qualification of materials used for processing and quantification of leachables and particulates reduces the risk of toxic substances entering the final process and interfering with the stability or purity of the drug product (more detail on quantification of leachables applicable to single- use bag systems is given in Section 6.1). The risk rating used to complete the FMEA shown in Table 6.14 and criticality of RPN can be visualised in the tables presented in Appendix 2. According to these ratings, most fill-finish operations are critical to product purity and quality (RPN >216). As a result, strict control systems should be put in place to reduce risk of batch/product rejection derived from serious non- compliance. In addition, as part of the continuous improvement of the process and risk management described in ICH Q9, the data collected during fill-finish operations should be reviewed continuously. These data focus on: manufacturing records, volume control records (recorded during the filling process), reports on equipment qualification, equipment manuals, calibration reports of pumps and stirrer tanks, reports of product composition, analysis methods, stability studies, deviations and complaints [21].

6.4.3 Qualification of fill-finish operations

Fill-finish PQ will carry out all operations within the fill finish process under normal operating conditions, these operations concern final drug product formulation,

Table 6.14: FMEA of a fill-finish process.

Risk	Process step	Failure mode	Effect of failure	S	Potential causes	O	Current controls	D	RPN
Contamination									
Non-sterile materials (e.g., vials, caps, tubing, WFI) used for processing	Sterilisation of materials used for processing	Inadequate sterilisation process (irradiation, autoclaving)	Contamination present in materials	8	Failure of sterilisation cycle	3	Steriliser parameters checked after process Quality assurance approval of sterilised materials before use / release	2	48
Contaminated materials enter cleanroom	Materials transfered to cleanroom	Inadequate cleaning process Inappropriate PPE	Contamination present in materials and/ or operator	4	SOP not followed Operator error Cleaning material out of date PPE faulty batch	2	Training of operators Gowning of operators Materials approved before use/release (e.g., cleaning, gloves)	3	24
Contaminated materials or cleanroom are used in processing	Cleaning and sanitation of materials and cleanroom	Inadequate cleaning process	Contamination present in materials or cleanroom	4	Inappropriate cleaning validation of cleaning agents Operator error (e.g., concentration, exposure time) Cleaning material out of date, faulty batch	2	Validation of cleaning materials and process post-sanitation EM done before cleanroom use/release	1	8
Contamination from operator and/or cleanroom	Post-room release Filling process	Weak area positive pressure	Contamination present in cleanroom	6	Failure in HVAC system	3	Alarms Pressure charts	2	36
	Material preparation & aseptic line set-up	Operator poor aseptic technique Operator error Inappropriate PPE System leaks	Contamination present in materials, cleanroom or operator	5	Faulty tubing line batch Faulty PPE batch	3	EM and personnel controls (e.g., finger plates, gowning) Monitoring (pressure charts) Training	3	45

(continued)

Table 6.14 (continued)

Risk	Process step	Failure mode	Effect of failure	S	Potential causes	O	Current controls	D	RPN
Contamination from system leaks	Filling process	Product contamination due to system leaks	Loss of product containment	5	Filter or tubing blockage	6	Verification of connections and leak test before start of filling process	3	90
Operational failure Incorrect Filling volume	Filling process	Inappropriate dosing delivery due to incorrect calibration or air bubbles in suspension of dosing pump	Incorrect amount delivered to final container / Discard solution for not meeting specification	5	Peristaltic pump stops or speed is higher/lower than required / Balance malfunction / Operator error	4	Calibration of fill pump and balance, as well as verification of correct fill amount before starting operation	4	80
	Filling process	Difference in sedimentation profile of product / Product dispensed in inappropriate format	Product settles in a specific area of the filling system and concentration uniformity of final filled product is modified / Discard solution for not meeting specification	5	Inadequate mixing speed and/ or time / Stoppage of operation / Filter or tubing blockage	4	Visual check of mixing speed/time / Install low speed stirrer alarm / Install system stoppage alarm / Install system automatically discard or manually discard product in-line / In-line device to measure suspension concentration	4	80
Purity and stability of the product Particulates from plastic material	Filling process	Inappropriate batch material supplied/used for processing	Particulates released from filter, bag, tubing or final container	8	May affect the purity of the final product / Potential toxic to patient	4	Materials qualified and released before processing / Visual inspection of final filled product	8	256

					S		O	D	RPN
Leachables from plastic material	Filling process	Inappropriate batch material supplied/used for processing	Leachables present in product contacting material	May affect the purity and stability of the final product. Potentially toxic to the patient	8	Materials qualified and released before processing	4	8	256
Changes in product profile	Filling process	Inappropriate dispensing speed/ pressure. Inappropriate tank sti.rrer speed	Too much shear affects/ modifies the product profile	Changes to final product (degradation, precipitation)	8	Procedures in place. Operator training	4	8	256

D: Detection

O: Occurrence

PPE: Personal protective equipment

S: Severity

sterilising filtration and filling into final container. As stated in the risk-assessment section, the most critical parameters for disposable systems undertaking final fill-finish is the guarantee of product sterility and that the right quantity is dispensed into final containers.

Validation of aseptic filling operations usually starts with media fills ('process simulations' – see Section 5.2.2), which includes using sterile tryptic soy broth growth media instead of the product to simulate filling operation (fill rates, container numbers, duration, environment, operators, equipment) under normal operating conditions. The dispensed fill quantities (weights/volumes) are recorded. Filled containers are evaluated for sterility, integrity and accuracy of fill quantity (weights/volumes). Media fill demonstrates the capability to produce sterile drug product reliably and robustly and to qualify personnel for aseptic processing. Further details regarding qualification of aseptic processes can be found in the literature [22] and FDA guidelines [14].

Besides qualification of normal operating conditions and filling accuracy, leachables and formation of particulates should be determined during processing of the drug product. Temperature, light, prolonged exposure to plastic surfaces and exposure to shear and air–liquid interfaces may affect the stability, quality and purity of the product during formulation and filling. Mixing during formulation, pumping during transfer and the filling process using a filling needle, for example, may create shear inducing air–liquid interfaces that could in turn cause degradation or aggregation of the product, and introduce particulates that can interact further with the API. Stability and purity of the API can also be affected by substances that migrate from the manufacturing equipment or final packaging container [23, 24]. Preliminary assessment of these substances is possible under a worse-case scenario. Small-scale characterisation to quantify and assess the impact of these substances upon the drug product is also possible. Any qualification will, however, require testing under normal operating conditions and with 'real' API stream. More details concerning extractables and leachables can be found in Section 6.1.

During PQ, samples must be taken over time to quantify leachables and the number of particles that may be produced during the filling process and which are derived from the sterilising filter, pumps, mixing tank, and pinch valve. Such tests are done with actual API under process conditions to prove that the product and process conditions do not result in an increase of particulates or leachable substances. Final product within the final container closures should also be tested for leachables throughout the duration of storage and under conditions of storage and transportation. Figure 6.12 presents a sequence of activities for formulation and fill-finish process, with critical activities such as sterile filtration, filling and storage highlighted. Potential expiry dates and sampling regimens to qualify these critical-operation hold times are proposed with a filtered drug product held for ≤72 h, and sterile components used for filling stored for ≤60 days. An expiry date of the final drug product of 36 months is considered. Hold study time points depend on risk

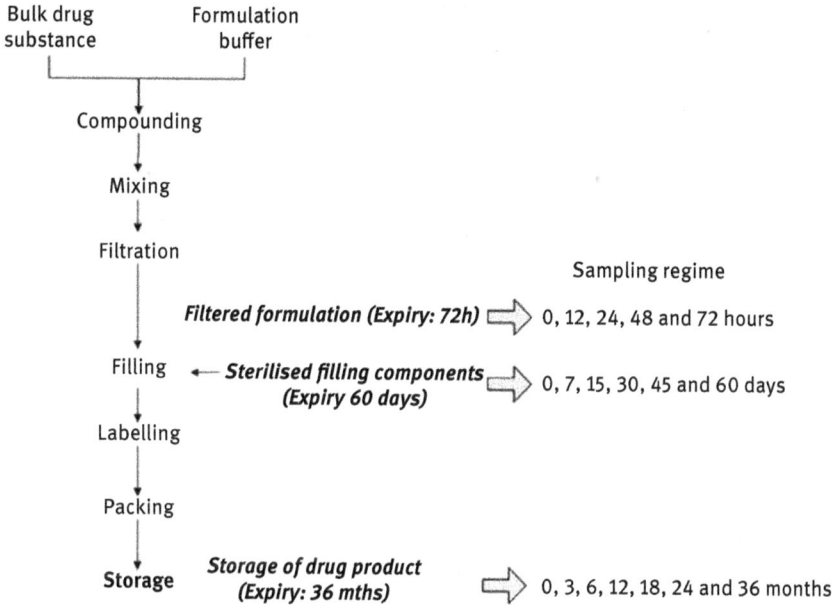

Figure 6.12: Fill-finish block flow diagram with hold study time points and sampling regimens.

assessment and (internal or external) capability for analysis but have been proposed to be every 3–6 months, 7 days or 12 h (including sampling at start of analyses).

Bacteria challenge tests undertaken with *Brevundimonas diminuta* according to HIMA/ ASTM F838-05 guidelines should be done in sterilising-grade filters as described in the FDA guideline for sterile drug products produced by aseptic processing [8]. This test, as well as the adsorptive properties of the filter, leachables and particulate formation should be done using the actual drug product.

References

[1] A. Sette and M. Barbaroux, *Biopharm International*, November 2006.
[2] Thermofisher Scientific, *Technical Paper*, 2011, **001**, Rev. 2, 1.
[3] Bio-Process Systems Alliance Guidelines and Standards Committee, *BioProcess International*, 2007, **5**, 4, 52.
[4] USP <1664.1>, First Supplement to USP 38-NF 33, 7193–7200.
[5] Bio-Process Systems Alliance Guidelines and Standards Committee, *BioProcess International*, 2008, **6**, 1, 44.
[6] USP <1663>, First Supplement to USP 38-NF 33, 7181–7193.
[7] E. Jurkiewicz, U. Husemann, G. Greller, M. Barbaroux and C. Fenge, *AIChemE*, 2014, **30**, 5, 1171.
[8] DeCheme Biotechnologie in *Recommendations for Leachables, Studies: Standardised Cell Culture Test for the Early Identification of Critical Films*, DeCheme Biotechnologie, Frankfurt, Germany, January 2014.

[9]   M. Hammond, H. Nunn, G. Rogers, H. Lee, A. Marghitoiv, L. Perez, Y. Nashed-Sameul, C. Anderson, M. Vandiver and S. Kline, *PDA Journal of Pharmaceutical Science and Technology*, 2013, **67**, 1, 123.

[10] Bio-Process Systems Alliance Guidelines and Standards Committee, Bioprocess interantional, 2007, 5, 11, 36.

[11] D. Chevaillier, M. Feuilloley, S. Genot, C. Lacaze, A. Laschi, Y. Legras and Uettwiller, *STP Pharma Pratiques*, 2013, **23**, 2, 1.

[12] W. Ding, G. Madsen, E. Mahajan, S. O'Connor and K. Wong, *Pharmaceutical Engineering*, 2014, **34**, 6. 1.

[13] *Guidance for Industry: Container–Closure Systems for Packaging Human Drugs and Biologics*, Department of Health and Human Services, Food and Drug Administration, Rockville, MD, USA, 1999.

[14] *Guidance for Industry Sterile Drug Products Produced by Aseptic Processing – Current Good Manufacturing Practice*, Department of Health and Human Services, Food and Drug Administration, Rockville, MD, USA, September 2004.

[15] D. Eibl and R. Eibl in *Single-use Technology in Biopharmaceutical Manufacture*, John Wiley & Sons, Inc., Hokoven, NJ, USA, 2011.

[16] De Wilde, T. Dreher, C. Zahnow, U. Husemann, G. Greller, T. Adams and C. Fenge, *BioProcess International*, 2014, **12**, 8s, 14.

[17] R. Reglin, S. Ruhl, J. Weyand, D. De Wilde, U. Husemann, G. Greller and C. Fenge, *BioProcess International*, 2014, **12**, 5, 53.

[18] A.G. Lopes, T. Selas, A.G. Hitchcock and D.C. Smith. *BioProcess International*, 2015, **13**, 9s, 1-8.

[19] A. Weber, U. Husemann, S. Chaussin, T. Adams, D. De Wilde, S. Gerighausen, G. Greller and C. Fenge. *BioProcess International*, 2013, **11**, 4, S6.

[20] B.A. Thorne, S. Waugh, T. Wilkie, J. Dunn and M. Labreck in *Implementation of SU TFF System in a cGMP Biomanufacturing Facility*, CMC Biologics, Bothell, WA, USA, November 2012.

[21] R. Diaz, G.F. Otero and C. Muzzio, *Pharmaceutical Engineering*, 2011, **31**, 3, 1.

[22] J.A. Allay and B.M. Belongia, *Pharmaceutical Technology*, 2nd March 2006.

[23] F. Jameel and S. Hershenson in *Formulation and Process Development Strategies for Manufacturing Biopharmaceuticals*, John Wiley & Sons, Inc., NJ, USA, 2010.

[24] J-P. Zambaux and J. Barry, *Bioprocess International*, April 2014.

Abbreviations

2D	Two-dimensional
3D	Three-dimensional
AET	Analytical evaluation threshold
API	Active pharmaceutical ingredient
ASTM	American Society for Testing and Materials
BOM	Bill of materials
BPSA	Bio-Process System Alliance
BSE	Bovine spongiform encephalopathy
CE	European conformity
cGMP	Current good manufacturing practices
CIP	Clean-in-place
COGs	Cost of goods sold
CPP	Critical process parameters
CQA	Critical quality attribute(s)
DF	Diafiltration
DO	Dissolved oxygen
DSP	Downstream processing
EM	Environmental monitoring
EtOH	Ethanol
EVA	Ethylene vinyl acetate
EVOH	Ethylene vinyl alcohol
FAT	Factory acceptance test
FDA	Food and Drug Administration
FMEA	Failure mode effects analysis
FS	Functional specification(s)
GEP	Good engineering practices
GMP	Good manufacturing practice(s)
HACCP	Hazard analysis and critical control point
HAZOP	Hazard and operability study
HIMA	Health Industry Manufacturers Association
HVAC	Heating ventilation air conditioning
ICH	International Conference on Harmonisation
ID	Internal diameter
IQ	Installation qualification
ISO	International Organization for Standardization
ISPE	International Society of Pharmaceutical Engineering
ISTA	International Safe Transit Association
MF	Microfiltration
MSAT	Manufacturing science and technology
OD	Outer diameter
OQ	Operational qualification
P&ID	Piping and instrumentation diagram
PA	Polyamide
PFD	Process flow diagram
PPE	Personal protective equipment
PQ	Performance qualification
PRA	Preliminary risk assessment

https://doi.org/10.1515/9783110640588-007

PW	Purified water
QA	Quality assurance
QC	Quality control
QMS	Quality management system
QRM	Quality risk management
R&D	Research and development
RPN	Risk priority number
SAT	Site acceptance testing
SIP	Steam-in-place
SME	Subject matter experts
SOP	Standard operating procedures
SUB	Single-use bioreactor(s)
SUT	Single-use technology
SWOT	Strength, weakness, opportunities and threats (analysis)
TFF	Tangential-flow filtration
TMP	Transmembrane pressure
TOC	Total organic carbon
TSE	Transmissible spongiform encephalopathy
UF	Ultrafiltration
UP	Upstream processing
URS	User requirements specification(s)
USP	Upstream processing
WFI	Water-for-injection

Appendix 1 Scoring Tables

Table A1.1: Risk-assessment scores according to risk factor.

Risk factor	Parameters	Score
Product contact		
Direct	–	14
Indirect	–	3
None	–	1
Location in process		
Final formulation	–	14
Downstream	–	6
Upstream	–	2
Temperature or contact time		
High	>70 °C or >24 h	10
Medium	37–70 °C or 12–24 h	6
Low	2–37 °C or <12 h	3
Surface area		
High	>0.6 m^2	10
Medium	0.05–0.6 m^2	6
Low	0.0005–0.05 m^2	3
Pre-treatment		
No flush before use	–	10
Flushed before use	–	6

https://doi.org/10.1515/9783110640588-008

Appendix 2 Risk Rating and Priority Number

Table A2.1: Rating for severity, occurrence and detection.

Risk category		Score	Definition of risk
Severity	Critical	8	Serious non-compliance; probable serious harm to patient; serious impact on yield or production capability; probable rejection of batch/product.
	High	6	Major non-compliance; probable impact on patient; high impact on yield or production capability; probable rework/re-processing of batch/product.
	Medium	4	Significant non-compliance; possible impact on patient; moderate impact on yield or production capability.
	Low	2	Minor non-compliance; no possible impact on patient, yield or production capability.
Occurrence	High	8	Highly probable to occur. Expected to occur >50% of the time.
	Medium	6	Improbable to occur. Expected to occur 10–50% of the time.
	Low	4	Improbable to occur. Expected to occur <10% of the time.
	Extremely low	2	Highly improbable to occur. Expected to occur <1% of the time.
Detection	High	2	Control system in place has a high probability of detecting failure mode (e.g., real-time monitoring temperature, pH, dissolved oxygen, mixing rate, times).
	Medium	4	Control system in place could detect failure mode (e.g., filter integrity test).
	Low	6	Control system in place has a low probability of detecting failure mode (e.g., visual detection of seals in culture bag or product hold in system, change in contamination profile of culture bag).
	Non-existent	8	There is no control system for detecting failure mode (e.g., sterility failure in sterile product fills).

Table A2.2: Criticality range of risk priority number.

Criticality	Risk priority number (RPN)
Critical	>216
Major	>40 and <216
Minor	<40

Index

https://doi.org/10.1515/9783110640588-009